Biology
Workbook
FOR
DUMMIES®

by René Fester Kratz, PhD

WILEY

John Wiley & Sons, Inc.

Biology Workbook For Dummies®

Published by
John Wiley & Sons, Inc.
111 River St.
Hoboken, NJ 07030-5774
www.wiley.com

WILEY

About the Author

René Fester Kratz, PhD, grew up near the ocean in Rhode Island. From a young age, she wanted to be a teacher (because she loved her teachers at school) and a biologist (because her dad was one). She graduated from Warwick Veterans Memorial High School and went on to major in biology at Boston University. As a freshman and sophomore at BU, René got excited by subjects other than biology and even considered changing her major. Then, she met and studied under Lynn Margulis, who reignited René's love of biology and introduced her to the world of microbes. René graduated with a BA in biology from BU and went on to earn an MS and a PhD in botany from the University of Washington. At UW, René studied reproductive onset in *Acetabularia acetabulum,* a marine green alga that grows as single cells big enough to pick up with your fingers. When they enter reproduction, the cells of *A. acetabulum* form a flat disk or cup-shaped structure at the top, earning the alga the nickname of the "mermaid's wine glass."

René currently teaches biology and general science classes at Everett Community College in Everett, Washington. She spends most of her time introducing students to the wonders of cells and microbes as she teaches cellular biology and microbiology. René also has a strong interest in science education and science literacy for everyone. As a member of the North Cascades and Olympic Science Partnership, she helped create inquiry-based science courses for future teachers that are based on research on human learning. She loves teaching these courses because they make science accessible for all kinds of people. In the summer, René enjoys working with K–12 teachers on the improvement of science education in the public schools. She also enjoys writing about science and is the author of *Molecular & Cell Biology For Dummies, Biology For Dummies* 2nd edition, *Botany For Dummies* (all published by Wiley), and *E–Z Microbiology,* 2nd edition (Barron's Educational Press).

René loves living in the Pacific Northwest because she is near the ocean and her daffodils start blooming in February (when her family back East is still shoveling snow). She doesn't mind the rain and thinks the San Juan Islands are one of the most beautiful places on Earth. Her husband, two sons, and two very bad dogs help her remember what is truly important, and her "sisters" help keep her sane. René loves to scrapbook, stitch, garden, and read.

Dedication

Happy 70th birthday to my mom, Annette — without your support and encouragement, I wouldn't be where I am today. Love you always.

Author's Acknowledgments

Thanks to Matt Wagner of Fresh Books, Inc., for helping me find the opportunity to work on this book. And thanks to all the great people at Wiley who made it happen: my project editor, Heike Baird; the executive editor, Lindsay Lefevere, who helped get me started on the project; Alicia South, who coordinated the art; and Kimberly Lyle-Ippolito and Allison Thomas, my technical reviewers. Thanks also to Nikki Gee, the project coordinator.

On the home front, thanks to my husband, Dan, and my sons, Hueston and Dashiel, for all their love and support.

Publisher's Acknowledgments

We're proud of this book; please send us your comments at http://dummies.custhelp.com. For other comments, please contact our Customer Care Department within the U.S. at 877-762-2974, outside the U.S. at 317-572-3993, or fax 317-572-4002.

Some of the people who helped bring this book to market include the following:

Acquisitions, Editorial, and Vertical Websites

Project Editor: Heike Baird

Executive Editor: Lindsay Sandman Lefevere

Copy Editor: Todd Lothery

Assistant Editor: David Lutton

Editorial Program Coordinator: Joe Niesen

Technical Editors: Kimberly Lyle-Ippolito, Allison Thomas

Senior Editorial Manager: Jennifer Ehrlich

Editorial Assistants: Rachelle S. Amick, Alexa Koschier

Art Coordinator: Alicia B. South

Cover Photos: © iStockphoto.com / Michał Rózewski

Cartoons: Rich Tennant (www.the5thwave.com)

Composition Services

Project Coordinator: Nikki Gee

Layout and Graphics: Carrie A. Cesavice, Joyce Haughey, Jennifer Mayberry, Christin Swinford

Proofreaders: John Greenough, Mildred Rosenzweig

Indexer: Palmer Publishing Services

Illustrator: Kathryn Born

Special Help: Medhane Cumbay

Publishing and Editorial for Consumer Dummies

    Kathleen Nebenhaus, Vice President and Executive Publisher

    Kristin Ferguson-Wagstaffe, Product Development Director

    Ensley Eikenburg, Associate Publisher, Travel

    Kelly Regan, Editorial Director, Travel

Publishing for Technology Dummies

    Andy Cummings, Vice President and Publisher

Composition Services

    Debbie Stailey, Director of Composition Services

Contents at a Glance

Table of Contents

Introduction

●●

*L*iving things are all around you, from the bacteria that live on your skin to the green plants that cover the land to the majestic blue whales that swim through the ocean. You're aware of many of these forms of life, but have you ever taken a look at the single-celled creatures in a drop of pond water? Or thought about the many different ways your life depends on the actions of plants and bacteria? The journey to discover more about the living world around you is at the heart of biology.

A living organism can be as simple as a single cell or as complex as a human being, but no matter how different they may seem, all living things on Earth have fundamental similarities: They're made of cells that contain DNA, and they all grow, move, get energy, use raw materials, make waste, and reproduce. These similarities among all living things illustrate how all life on Earth is part of the same big family tree. And the differences in how each type of living thing achieves these same goals — well, the differences are what make biology fascinating.

Beyond increasing your appreciation for other kinds of life, the science of biology can help you understand your place in the living world. At first glance, you may think that people can do everything for themselves — get food from the grocery store, build their homes, and make their clothes — but a closer look quickly shows how dependent people are on the rest of the living world. People depend on plants and green bacteria to make the food that supports food chains that include agricultural species. These green organisms also make the oxygen that people need to sustain life. Bacteria and fungi in the soil break down dead organisms, recycling matter so that other living things can reuse it. Clearly, people can't survive on planet Earth alone.

I hope that you enjoy your exploration of the living world and come to appreciate the marvelous diversity of life on Earth. I also hope that the information in this book improves your performance in biology class, specifically (gulp!) your exams. This truly is a living planet, and the more that people understand the connections among living things, the better choices they'll make about the future of the world.

About This Book

Biology Workbook For Dummies is designed to help supplement your learning in a biology class or to use as a companion for your self-guided exploration of biology using *Biology For Dummies,* 2nd edition (Wiley). This workbook isn't intended to replace a textbook but rather to highlight essential information in an easy-to-understand format and quiz you on it. I provide many straightforward lists of the fundamentals you need to know about the various subjects you'd typically encounter in a biology class, along with problems on which you can practice and reinforce your understanding. I provide answers to all the practice questions and include explanations of why some answers are right or wrong.

If you're taking biology, your instructor may present material in a different order than the organization I use here, so be sure to take advantage of both the table of contents and the index to navigate where you need to go.

Conventions Used in This Book

In order to explain topics as clearly as possible, I keep scientific jargon to a minimum and present information in a straightforward, linear style. I break dense information into main concepts and divide complicated processes into series of steps.

To help you find your way through the subjects in this book, I use the following style conventions:

- ✔ I use *italic* for emphasis and to highlight new words or terms that I define in the text.
- ✔ I use **boldface** to indicate key words in bulleted lists and the action parts of numbered steps.
- ✔ I use `monofont` for web addresses so they're easy to recognize.
- ✔ When this book was printed, some web addresses may have needed to break across two lines of text. If that happened, know that I haven't put in any extra characters (such as hyphens) to indicate the break. So when using one of these web addresses, type in exactly what you see and ignore the line break.

Foolish Assumptions

As I wrote this book, I tried to imagine who you might be and what you might need to understand biology, and here are some assumptions I made:

- ✔ You may be a high school student taking biology and maybe preparing for an advanced placement test or a college entrance examination. For you, I've tried to extract the essentials about each subject and organize them into short lists that are easier to study than long paragraphs. I've also written problems for you to practice on and given you links to websites with great animations and tutorials.

- ✔ You may be a college student who isn't a science major but is taking a biology class to help fulfill your degree requirements. For you, I've tried to get the main ideas across with as little scientific jargon as possible. If you find that you get overwhelmed in your biology class, try reading a section in this workbook before you go to class to hear a lecture about the topic. That way, you'll have some of the big ideas in your mind before your instructor starts adding all the details. Also, many non-science students are a little shocked to find that their usual study habits don't work well for a science class, so be sure to read my tips in Chapter 20 on how to get an A in biology.

- ✔ You may be someone who just wants to know a little bit more about the living world around you. You may have picked up *Biology For Dummies,* 2nd edition, or some other biology text because you want to take a deeper look at the living world. For you, this workbook will make a nice companion and give you a chance to test yourself on the practice problems to see how well you're learning the information.

Whatever your reason for picking up this book, I've done my best to explain the topics of biology simply and effectively and to create some challenging practice problems to help improve your learning. I hope you find this workbook helpful.

How This Book Is Organized

I've arranged this book to follow the order of topics in many biology textbooks, with a few minor differences. Like all *For Dummies* books, each chapter is self-contained, so you can pick up the book whenever you need it and jump into the topic you're working on. After I explain a subject, I use that information in later topics. So if you don't read the book in order, you may occasionally have to refer back to an earlier section for some background information. When that's the case, I refer you to the appropriate section or chapter.

Part 1: Getting the Basics

Biology is the study of life, but as I'm sure you know, life is complex. To simplify it, I break the all-encompassing subject of biology into smaller, more palatable chunks. To start, I explain the way that scientists study biology. This *scientific method* holds not only for biology but also for chemistry, psychology, physics, geology, and other sciences and social sciences. Knowing how scientists conduct, challenge, check, and recheck research makes it easier to appreciate the value of scientific information.

In this part, I spotlight the basic unit of life: the cell. Every organism — whether it's a human, a dog, a flower, a strep throat bacterium, or an amoeba — has at least one cell; most have millions. After you have a grasp of how cells are the powerhouses of bodies, I review the types of molecules that are important to their functioning. Included in this first part is the often-dreaded but oh-so-necessary review of basic chemistry. To learn biology, you must understand some basic principles of how chemicals function. After all, the bodies of every living organism are big sacs of chemicals. Chemical reactions generate every process that occurs in your body, such as those that occur during the metabolic processes in plant and animal cells. So in this part you delve into topics such as enzymes, energy transfer, and how plants make food from scratch, using just carbon dioxide and water!

Part II: Creating the Future with Cell Division and Genetics

Cells reproduce, giving rise to other cells. Sometimes cells make exact copies of themselves in order to repair, grow, or produce offspring that are genetically identical to the parent. Some organisms mix it up a little by engaging in sexual reproduction, creating offspring that have combinations of genes that are different from those of their parents.

But whether organisms reproduce asexually or sexually, the parents' traits are visible in their offspring. Ducks make ducklings, and from little acorns mighty oaks do grow. Offspring inherit their traits from their parents because parents pass DNA to their offspring. DNA contains the blueprints for proteins that do the work in cells and thus determine an organism's characteristics. Biologists today are busy unraveling the mysteries of DNA, giving humans unprecedented power over the very stuff of life. This part walks you through all these topics and gives you plenty of chances to test yourself on what you've read.

Part III: Making Connections with Ecology and Evolution

All the amazingly diverse forms of life on Earth interact with one another. In this part, you become more aware of the living part of Earth as one big, interconnected ecosystem called the _biosphere_. Living things aren't just connected with one another today; they also have connections to the living things of the past. The science of evolution studies those relationships and uses them to understand present and future changes in the populations of living things on Earth today.

Part IV: Getting to Know the Human Body

Organisms respond to changes in their environment, trying to maintain their internal conditions within a range that supports life. Animals have many different systems that support this balance, which is called _homeostasis_. In this part, I explain most of the systems that support the structure and function of the human body and touch on how humans compare to other animals.

Part V: Going Green with Plant Biology

Our green neighbors are very quiet and sometimes get overlooked in the hustle and bustle of animal life. However, the importance of plants to life on Earth simply can't be overstated — they're the food makers, after all. Without plants (and green microbes), nobody else would have anything to eat! And when you take a good look (and you do in this part), plants are pretty interesting. Just like animals, they're made of cells and have systems to transport materials around their body and exchange matter and energy with their environment. Their structures are well suited to their lifestyle, and many plants are things of beauty. Just ask someone in your life who likes to garden! To test your green thumb, flip to the chapters in this part.

Part VI: The Part of Tens

This part contains two short chapters with lists of ten or so items. I give you tips for getting an A in biology and links for some websites that will help you do just that!

Icons Used in This Book

I use the familiar *For Dummies* icons here to help guide you and give you new insights as you read the material.

The text near this bull's-eye symbol may help you remember the facts being discussed or suggest a way to help you commit them to memory. Also, although you can learn most biological information on its own, some topics aren't clear unless the building blocks of information are stacked. In those instances, I provide info at this icon that I may have explained in an earlier chapter.

This icon marks sample problems that I've laid out step by step to help guide you through the solution. Reviewing these problems will help you answer similar problems in the practice problems or in those assigned by your biology instructor.

This icon serves to summon your memory. The information I spotlight here is info I think you should permanently store in your brain's biology file. If you want a quick review of biology, scan the book reading the text by these icons. No need for a chunky yellow highlighter.

The bomb icon marks ideas that commonly trip up students of biology. To move beyond these common misunderstandings, you need to confront them head on, and this icon helps you do that.

Where to Go from Here

With *Biology Workbook For Dummies,* you can start anywhere in the book that you want. If you're in the thick of a biology class and having trouble, jump right to the subject that's confusing you. If you're using the book as a companion to a biology class that's just beginning, the book follows the organization of most college classes, with one exception — most college classes work from the smallest to the largest, starting with molecules and then moving on to cells. I prefer to start with cells to give you a sense of context and then move on to the molecules. If you're reading this workbook for general interest, you'll probably find it best to begin at the beginning with the chapter on cells and then move on to whatever interests you next. Whatever your circumstance, the table of contents and index help you find the information you need.

I wrote this book with the non-scientist in mind. If you're taking your study of biology further and need more information, several other *For Dummies* books expand on the topics that I present in general terms here:

✔ My book *Molecular and Cellular Biology For Dummies* (Wiley) takes a deeper look into cells, basic cellular chemistry, metabolism, genetics, and the study of DNA.

✔ *Anatomy & Physiology For Dummies,* by Maggie Norris and Donna Rae Siegfried, presents more details on the structure and function of the human body, and *Anatomy & Physiology Workbook For Dummies,* by Janet Rae-Dupree and Pat DuPree, gives you lots of practice problems on the subject. Both of these books are published by Wiley.

✔ *Evolution For Dummies,* by Greg Krukonis and Tracy Barr (Wiley), explores the topic of evolution more fully, looking at the evidence for evolution and the many mechanisms by which it occurs.

Best wishes from me to you as you begin your exploration of life on Earth.

Part I
Getting the Basics

The 5th Wave — By Rich Tennant

BEING CHRONICALLY LATE, ANDY ALWAYS MISSED OUT ON EXAMINING THE REALLY FUN CELLS

CELL RESEARCH

"Sorry, Andy. All I got left are lung and gut cells. Take 'em or leave 'em."

In this part . . .

*B*iology is the scientific study of living things. Like all scientists, biologists use their five senses to ask questions about the natural world. Biologists follow a scientific method of asking questions, proposing answers, and then testing those answers through experimentation. This book starts off by explaining more about the scientific method and giving you some opportunities to practice using it.

All living things are made of cells, and cells are made of molecules. Living things with many cells, like humans, are organized into organ systems, organs, tissues, and cells. Cells are the smallest things that show all the properties of life, so this part gets you acquainted with cells so you can understand more about everything around you. I describe the structure of cells and the molecules that make them up, and then I explain how cells get the energy they need to function.

Chapter 1

Figuring Out the Scientific Method

● ●

In This Chapter

▶ Testing hypotheses using the scientific method

▶ Conducting scientific experiments the right way

▶ Distinguishing between hypotheses and theories

● ●

*B*iology is the branch of science that deals with living things. Biology wouldn't have gotten very far as a science if biologists hadn't used structured processes to conduct their research and hadn't communicated the results of that research with others. You can use what you learn in this chapter in your everyday life to take a closer look at the information that swirls all around you. Does that diet plan really work? What studies did they do? Ninety-seven percent of scientists agree that global warming is really happening. Why do they think that? What evidence are they looking at? This chapter introduces you to the methods that scientists (whether they're biologists, physicists, or chemists) use to investigate the world around them and helps you learn to analyze scientific experiments.

Developing Hypotheses

The true heart of science isn't a bunch of facts; it's the method that scientists use to gather those facts. Science is about exploring the natural world, making observations using the five senses and intellect, and attempting to make sense of those observations.

When scientists seek out, observe, and describe living things, they're engaging in *discovery science.* Scientists practice discovery science as they explore new environments, like the deep sea, describing the organisms they find there. As scientists study the natural world, they look for patterns and attempt to make sense of how things work. When a scientist proposes an untested explanation for how things work, the tentative explanation is called a *hypothesis.* When scientists test their understanding of the world through experimentation, they're engaging in *hypothesis-based science,* which usually calls for following some variation of a process called the *scientific method* (see the section "Practicing the Scientific Method" later in the chapter). For a hypothesis to be accepted by scientists, it must be *testable* or *falsifiable.* In other words, it must be an idea that you can support or reject by exploring the situation further and collecting observations using your five senses.

For example, let's say that you have a bird feeder in your backyard. You keep filling the feeder with birdseed, but every day when you get up, it's empty again. When you examine the feeder, you notice some scratches near the feeder hole that look like marks from animal claws, so you think that squirrels may be getting into your birdseed. So you take some wire screen and nail it over the feeder hole to reduce the size of the openings. After that, your birdseed lasts for days, and you observe birds eating at your bird feeder.

In my example, you took a scientific approach to solving your bird feeder problem.

✔ You made initial observations about your bird feeder constantly being emptied and further observed the scratch marks around the feeder hole.

✔ You came up with a hypothesis about the cause of the disappearing food: If a squirrel is stealing the food, then a smaller opening on the bird feeder would prevent that.

✔ You were able to test your hypothesis by making a change (creating a smaller opening) and then making new observations.

If an explanation isn't testable, it's not considered a scientific hypothesis.

In the bird feeder example, you may have thought, "Squirrels really enjoy annoying birds, and that's why they're stealing the food." This explanation relates to your observations, but unless you're an expert in reading squirrel emotions, it's not really something you can test and it wouldn't be considered a scientific hypothesis.

See if you can think like a scientist by answering these questions about observations and hypotheses:

1. Two scientists are mapping the locations of mushrooms in the Amazon rain forest. Thus, they're practicing

   a. Discovery science

   b. Hypothesis-based science

   c. Making observations

   d. Discovery science and making observations

2. One night as it gets dark, the scientists notice that some mushrooms glow in the dark. Which of the following would be a valid scientific hypothesis about this observation?

   a. The mushrooms glow because they're scared of the dark.

   b. The mushrooms glow to attract certain insects.

   c. The glowing mushrooms appear yellow-green in color.

Practicing the Scientific Method

Although the bird feeder and squirrel story from the preceding section is an everyday example, it illustrates the most important components of the scientific method. Scientists use the same procedure to make sense of the world whether they're studying squirrels in the backyard or the potential for life outside planet Earth, and that procedure is the scientific method.

The *scientific method* is basically a six-step plan that scientists follow while performing scientific experiments and writing up the results. By following the scientific method carefully, scientists make sure that their conclusions are based on observations and that other scientists can repeat their experiments. Here's the general process of the scientific method:

1. **First, make observations and come up with questions.**

   The scientific method starts by scientists noticing something and asking questions like "What's that?" or "How does it work?" — just like a child might when he sees something new, such as an earthworm wriggling in a puddle after a rainstorm.

2. **Then form a hypothesis.**

   Scientists form hypotheses using *inductive reasoning;* that is, they use specific observations to try and come up with general principles. Say, for example, a marine biologist is exploring a beach and finds a new worm-shaped creature he has never seen before. Using inductive reasoning, he may reach the hypothesis that the creature is some kind of worm because it's shaped like a worm.

3. **Next, make predictions and design experiments to test those ideas.**

   Predictions set up the framework for an experiment to test a hypothesis, and they're typically written as "if . . . then" statements. In the preceding worm example, the marine biologist predicts that if the creature is a worm, then its internal structures should look like those in other worms he has studied.

4. **Test the ideas through experimentation.**

   Scientists must design their experiments carefully to test just one idea at a time (I explain how to set up a good experiment in the "Designing Experiments" section, later in the chapter). As they conduct their experiments, scientists make observations using their five senses and record these observations as their *results* or *data*. Continuing with the worm example, the marine biologist tests his hypothesis by dissecting the wormlike creature, examining its internal parts carefully with the assistance of a microscope, and making detailed drawings of its internal structure.

   Any scientific experiment must have the ability to be duplicated because the "answer" the scientist comes up with (whether it supports or rejects the original hypothesis) can't become part of the scientific knowledge base unless other scientists can perform the exact same experiment and achieve the same results.

5. **Then make conclusions about the findings.**

   Scientists interpret the results of their experiments through *deductive reasoning,* using their specific observations to test their general hypothesis. When making deductive conclusions, scientists consider their original hypotheses and ask whether they could still be true in light of the new information gathered during the experiment. If so, the hypotheses can remain as possible explanations for how things work. If not, scientists reject the hypotheses and try to come up with alternate explanations (new hypotheses) that can explain what they've seen. In the earlier worm example, the marine biologist discovers that the internal structures of the wormlike creature look very similar to another type of worm he's familiar with. He can therefore conclude that the new animal is likely a relative of that other type of worm.

6. **Finally, communicate the conclusions with other scientists.**

   Communication is a huge part of science. Without it, discoveries wouldn't be passed on, and old conclusions wouldn't be tested with new experiments. When scientists complete some work, they write a paper that explains exactly what they did, what they saw, and what they concluded. Then they submit that paper to a scientific journal in their field. Scientists also present their work to other scientists at meetings, including those sponsored by scientific societies. In addition to sponsoring meetings, these societies support their respective disciplines by printing scientific journals and providing assistance to teachers and students in the field.

Continue testing your scientific thinking by answering these practice questions about the scientific method. Questions 3 through 5 refer to the following story:

Two scientists who are studying mushrooms in the Amazon rain forest discover a type of mushroom that glows in the dark. One of the scientists proposes that the mushrooms glow in order to attract a certain insect that will scatter the mushroom's reproductive spores. The scientists watch the mushrooms for several days, collecting samples of any insects that visit the mushrooms. Then, they take some of the mushrooms and insects back to their lab and test each type of insect to see whether it's attracted to the glowing mushroom. However, none of the insects shows any attraction to the glow. The scientists decide that the glow from the mushroom must have some other purpose than to attract any of the insects they collected.

3. Put yourself in the place of these scientists and write what you think they may have predicted for their experiment.

4. When the scientists decide that the glow from the mushroom has some other purpose than attracting any of the insects they collected, they're

　　a. Making an observation

　　b. Collecting data

　　c. Using inductive reasoning

　　d. Using deductive reasoning

5. Which of the following is an example of the type of data the scientists may have collected during their experiment?

　　a. The number of times a particular type of insect flew toward the glowing mushroom.

　　b. A comparison of the number of times an insect flew toward the glowing mushroom and away from the glowing mushroom.

　　c. The purpose of the glow may be to keep insects away from the mushroom.

　　d. The scientists ask their colleagues who work on glowing bacteria for information about what makes bacteria glow.

Designing Experiments

When a scientist designs an experiment to test her hypothesis, she tries to develop a plan that clearly shows the effect or importance of each factor tested by her experiment. Any factor that can be changed in an experiment is called a *variable*.

Three kinds of variables are especially important to consider when designing experiments:

　✔ **Experimental variables:** Also called *independent variables,* these are the factors you want to test or that are controlled by the researcher.

　✔ **Responding variables:** Also called *dependent variables,* these are the factors you measure. The dependent variable depends on the independent variable and is usually what ends up in your data table.

　✔ **Controlled variables:** These are any factors that you want to remain the same regardless of the changes in the experimental variables.

Scientific experiments help people answer questions about the natural world. To design an experiment,

1. **Make observations about something you're interested in and use inductive reasoning to come up with a hypothesis that seems like a good explanation or answer to your question.**

   For example, you're a runner who trains with a group of friends, and you have a hunch that loading up on pasta, which has lots of carbohydrates, gives you the energy you need to run faster the next day.

2. **Think about how to test your hypothesis.**

   One way to help focus your thinking is to create a prediction about your hypothesis using an "if . . . then" format. Translate that hunch into a proper hypothesis, which looks something like this: If a runner consumes large quantities of carbohydrates before a race, he'll run faster.

3. **Decide on your experimental treatment.**

   The condition or situation you alter in your experiment is your experimental (independent) variable. You can test your hypothesis by convincing half of your friends to eat lots of pasta the night before the race. Because the factor you want to test is the effect of eating pasta, pasta consumption is your experimental variable.

4. **Decide what to measure and how often to make measurements.**

   The changes you measure are your responding (dependent) variables. Race duration is your responding variable because you determine the effect of eating pasta by timing how long each person in your group takes to run the race.

5. **Create two groups of individuals for your experiment.**

   One group is your *experimental group* and the other is a *control group*.

   a. The experimental group receives the experimental treatment; in other words, you vary the one condition you want to test. In this case, you feed your friends pasta.

   b. The control group should be as similar as possible to your experimental group except that it doesn't receive the experimental treatment — so, no pasta for this group.

   For example, you convince half of your friends to eat a meal without pasta the night before the race. For the best results in your experiment, this control group should be as similar as possible to your experimental group so you can be pretty sure that any effect you see is due to the pasta and not some other factor. So ideally, both groups of your friends are about the same age, same gender, and same fitness level. They also eat about the same thing before the race, with the sole exception of the amount of pasta they eat at dinner. All the factors that could be different between your two groups (age, gender, fitness, and diet) but that you try to control to keep them the same are your *controlled variables*.

   Don't confuse controlled variables with the control groups. Controlled variables are the conditions you keep the same for all your groups, while the control group is the group of subjects in your experiment that you don't add any experimental variables to.

6. **Conduct your experiment.**

   Your friends eat their assigned meals the night before the race and then compete in the race the next day.

7. **Make your planned measurements and record them in a notebook.**

   Be sure to date all your observations. The observations you make are the *data* or *results* of your experiment.

   a. *Quantitative data* is numerical data like height, weight, and number of individuals that show a change. You can analyze quantitative data with statistics and present it in graphs.

   Scientists carefully record exact measurements from their experiments and present that data in graphs, tables, or charts. For this example, you average the race times for your friends in each of the two groups and present the information in a small table.

   b. *Qualitative data* is descriptive data like color, health, and happiness. You usually present qualitative data in paragraphs or tables.

   For your race experiment, you might ask your friends how they felt during the race: Did they have lots of energy? Did their energy level feel constant, or did they tire quickly?

8. **Analyze your data by comparing the differences between your experimental and control groups.**

   You can calculate the averages for numerical data and create graphs that illustrate the differences, if any, between your two groups.

   Your graph shows that your pasta-eating friends ran the marathon an average of two minutes faster than your friends who didn't eat pasta.

9. **Use deductive reasoning to decide whether your experiment supports or rejects your hypothesis and to compare your results with those of other scientists.**

   Because your pasta-eating friends ran faster, you may conclude that your hypothesis is supported and that eating pasta does in fact help marathon runners run faster races. You might also look at studies on other factors, like drinking enough water, and how they affect marathon speeds in order to compare the effect of your study to those of other scientists. If the best any other study did was decrease marathon times by 30 seconds and you decreased them by 2 whole minutes, you might conclude that your experimental variable — eating pasta — was more important than the variables tested in the other studies.

10. **Report your results.**

    Explain your original ideas and how you conducted your experiment, present your results, and explain your conclusions.

For a small study like the one I've used as an example, you might just report it informally by telling your friends or writing about it on your blog. But if you were an exercise researcher who conducted a large, well-designed study with lots of marathon runners, you'd write an article about your work and how it compared to the work of other researchers. Scientists in every field have their own special magazines, called *scientific journals*. You'd find a journal appropriate to your work, like *The International Journal of Exercise Science,* and submit your article to the editor. The editor would send your article out to other scientists or your peers in the field so that they could examine your work and decide whether it was good work that was worthy of publishing. Peer review is incredibly important to the process of science because it gives strength to scientific conclusions when others scientists can evaluate the same data and reach the same conclusion.

Scientific articles go through a process called *peer review* before they're published in scientific journals. During peer review, experts in the same field as the article's author examine the scientist's work to decide whether the experiments were conducted properly and whether the author's conclusions are valid based on the evidence collected.

Analyzing an experiment and really understanding experimental design is tough stuff. To help you understand, revisit the mushroom scientists again and take a closer look at their experiments. Questions 6 through 10 refer to the following story:

Two scientists want to test whether a glow-in-the-dark mushroom glows in order to attract insects. To test their idea, they set up an experiment. First, they build three identical chambers that are completely dark. In one chamber, they put a glowing mushroom. In another chamber, they put a light that glows the exact same color as the mushroom. In the third chamber, they put a mushroom that's a close relative to the glowing mushroom but that doesn't glow. The scientists put the same species of insect into each chamber and observe whether the insect flies toward or away from the mushrooms or the light. They repeat this procedure several times in each chamber, using new insects of the same species each time.

6. What's the experimental variable in the scientists' experiment?

   a. The type of insect used

   b. The size of the chamber

   c. The object placed in the chamber with the insect

   d. The direction that the insect flies when placed in the chamber

7. What's an example of a controlled variable in the scientists' experiment?

   a. The type of insect used

   b. The light bulb placed in the chamber with the insect

   c. The type of mushroom placed in the chamber with the insect

   d. The direction that the insect flies when placed in the chamber

8. What's the control group in the scientists' experiment?

   a. The type of insect used

   b. The insects placed in the chamber with the glowing mushroom

   c. The insects placed in the chamber with the nonglowing mushroom

   d. The insects placed in the chamber with the glowing light bulb

9. Which of the following is an example of qualitative data that the scientists may have collected during their experiment?

   a. The number of times the insects flew toward the test object

   b. The number of times the insects flew away from the test object

   c. The pattern of the insects' flight (straight lines versus wandering)

   d. The speed at which the insects flew

10. What's the responding variable in the scientists' experiment?

a. The type of insect used

b. The light bulb placed in the chamber with the insect

c. The type of mushroom placed in the chamber with the insect

d. The direction that the insect flies when placed in the chamber

Making an Experiment Count

A scientist may consider all the variables carefully and design a good experiment, but in order for an experiment to be valid and significant to the scientific community, it must meet these standards:

✔ **Sample size:** The number of individuals that receive each treatment in an experiment is your *sample size*. To make any kind of scientific research valid, the sample size has to be large. If only four of your friends participate in the pasta experiment in the preceding "Designing Experiments" section, you'd have to conduct your experiment again on much larger groups of runners with hundreds of people per group before you could proudly proclaim that consuming large quantities of carbohydrates before a race helps marathon runners improve their speed. The larger the sample size, the more valid the conclusions from an experiment.

✔ **Replicates:** The number of times you repeat the entire experiment or the number of groups you have in each treatment category are your *replicates*. Suppose you have 60 marathon-running friends and you break them into six groups of 10 runners each. Three groups eat pasta and three groups don't, so you have three replicates of each treatment. (Your total sample size is therefore 30 for each treatment.)

✔ **Statistical significance:** The mathematical measure of an experiment's validity is referred to as *statistical significance*. Scientists analyze their data with statistics to determine whether the differences between groups are significant. If you perform an experiment repeatedly and the results are within a narrow margin, the results are said to be significant. In your experiment, if the race times for your friends were very similar within each group, and all your pasta-eating friends ran the race two minutes faster than your non-pasta-eating friends, then that two-minute difference actually means something. But what if some pasta-eating friends ran slower than non-pasta-eating friends, and one or two really fast friends in the pasta group lowered that group's overall average? Then you may question whether the two minutes is really significant or whether your two fastest friends just got put in the pasta group randomly.

✔ **Error:** Science is measured by people, and people make mistakes, which is why scientists always include a statement of possible sources of error when they report the results of their experiments. Consider the possible errors in your experiment. What if you didn't specify anything about the content of the meals without pasta to your non-pasta-eating friends? After the race, you may find out that some of your friends ate large amounts of other sources of carbohydrates, such as rice or bread. Because your hypothesis is about the effect of carbohydrate consumption on marathon running, a few friends eating rice or bread would represent a source of error in your experiment.

Whether a scientist's initial hypothesis is right or wrong isn't as important as whether he sets up well-designed, repeatable experiments that provide necessary information to advance the frontiers of scientific knowledge.

Practice identifying sample size and number of replicates by answering the following questions. Questions 11 and 12 refer to the following story:

Recall from the preceding section that the scientists who were testing glowing mushrooms built three identical chambers. In one chamber, they put a glowing mushroom. In another chamber, they put a light that glows the exact same color as the mushroom. In the third chamber, they put a mushroom that's a close relative to the glowing mushroom but that doesn't glow. To determine whether the glowing color of the mushroom was attracting the insects, the scientists put the same species of insect into each chamber and observed whether the insects flew toward or away from the mushrooms or the light. On Monday, the scientists tested ten insects in each chamber and recorded their results. On Tuesday, they returned to the lab and tested another ten insects in each chamber.

11. What's the sample size for each experimental variable for the scientists' experiment?

   a. 3

   b. 10

   c. 30

   d. 60

12. How many replicates did the scientists perform?

   a. 2

   b. 3

   c. 10

   d. 60

Building Theories

The knowledge that scientists gather continues to grow and even change slightly over time. Scientists are continually poking and prodding at ideas, always trying to get closer to "the truth." They try to keep their minds open to new ideas and to remain willing to retest old ideas with new technology. In a way, science is an adventure, with scientists as the explorers trying to map new territory. As scientists move into new areas, they create new maps. And as new tools become available, scientists refine old maps, making them more accurate. When a "map" of some particular idea isn't quite finished, scientists may argue over the details of how it should be drawn. Scientists encourage debate over ideas because it pushes them to test their ideas and ultimately adds to the strength of scientific knowledge. The goal in science isn't to win an argument but rather to find the explanation that best fits all the observable data.

Theories are scientific explanations based on a large body of evidence that usually comes from the efforts of many different scientists. Although hypotheses are also explanations, they're based on initial observations and haven't yet been subjected to rigorous testing.

The way scientists use the word *theory* is very different from the way most people use the word. In everyday language, people use the word *theory* to mean a guess, but in scientific language, a theory is as close as scientists get to saying an idea is true. Because scientists never stop exploring and adding to their knowledge, they tend to avoid saying that something is absolutely true or fact. They like to leave a little room open for ideas to be modified or expanded. However, theories are so well-supported by evidence that they rarely undergo big changes. Usually, changes to theories are more along the lines of minor modifications. Scientific theories that most people accept as true include the germ theory of disease (microbes like bacteria can cause disease), the theory of plate tectonics (Earth's surface is made up of separate plates that float upon the Earth's mantle), and the cell theory (all living things are made of cells). The theory of evolution by natural selection is also extremely well supported by many lines of scientific investigation and accepted by most scientists as true.

Can you tell the difference between a hypothesis and a theory? Try the following questions to find out:

13. Two scientists who are studying glowing mushrooms notice that the mushrooms glow the exact same color as some glowing bacteria in the lab next door. One of the scientists says, "I think the glowing chemical inside the mushroom is the same as the glowing chemical inside the bacteria." What type of statement is the scientist making?

a. Scientific (testable) hypothesis

b. Nonscientific guess

c. Scientific theory

d. Deductive reasoning

14. How does the statement made by the scientist in Question 13 compare to a scientific theory? If you think it's a scientific theory, explain why. If you don't think it's a scientific theory, explain how the statement is different from a scientific theory.

Answers to Questions on the Scientific Method

The following are answers to the practice questions presented in this chapter.

 The answer is **d. Discovery science and making observations.**

The scientists are practicing discovery science because they're describing something rather than doing an experiment. They're making observations because they're using one of their five senses to see the mushrooms.

2 The answer is **b. The mushrooms glow to attract certain insects.**

A hypothesis is a tentative explanation that can be tested using the five senses. The idea that mushrooms are scared of the dark is a tentative explanation, but it's not testable using the five senses. The color of the mushrooms is a simple observation, not an explanation of how or why the mushrooms glow.

3 You should have written something like, "**If the mushrooms glow in order to attract insects, then some insects should be attracted to the glowing mushrooms.**" Remember that predictions are usually written as "if . . . then" statements.

4 The answer is **d. Using deductive reasoning.**

The key word in this question is "decide." The fact that the scientists are deciding something tells you that they've moved beyond making observations and collecting data. Now they're using their brains to make decisions, or conclusions. Their decision is based on the data they've collected, so they're using deductive reasoning.

5 The answer is **a. The number of times a particular type of insect flew toward the glowing mushroom.**

To look at the other possible answers, comparing data requires thinking, so that's part of decision-making. Deciding on a new possible explanation for the glow's purpose is forming a new hypothesis. And talking to colleagues represents communication and collaboration among scientists.

6 The answer is **c. The object placed in the chamber with the insect.**

The experimental variable is the factor that the scientists were testing and therefore changing between treatments.

7 The answer is **a. The type of insect used.**

Controlled variables are factors that the scientists try to keep the same between treatments so they don't affect the outcome of the experiment.

8 The answer is **b. The insects placed in the chamber with the glowing mushroom.**

The control group is a group of subjects that don't receive an experimental treatment. It's often a group that's exposed to what's considered normal conditions. The scientists would compare the results of treatments **c** and **d** with **b** to figure out the effect of type of mushroom versus glow on the insects.

9 The answer is **c. The pattern of the insects' flight (straight lines versus wandering).**

You can measure, or quantify, factors like number of times and speed, so they're examples of quantitative data.

10 The answer is **d. The direction that the insect flew when placed in the chamber.**

The responding variable is what scientists measure.

11 The answer is **b. 10.**

The sample size of an experiment is the number of individuals in a treatment. Each day, the scientists tested 10 individual flies in each chamber.

12 The answer is **a. 2.**

The number of replicates is how many times the scientists repeated their experiment. They did a complete run of their experiment on Monday and again on Tuesday.

13 The answer is **a. Scientific (testable) hypothesis.**

The scientist is proposing an idea based on an observation but not based on detailed testing of the idea.

14 The scientist's statement in Question 13 is a tentative or proposed explanation. To develop a theory, the scientists would have to study glowing mushrooms in a great amount of detail and compare their research to that of other scientists. For example, the scientists studying glowing mushrooms may be interested in understanding why mushrooms glow, how they glow, and whether the ability to glow has evolved many times in living things or just once. If they study their mushrooms and combine their information with that of other scientists to build a larger understanding of how and why living things glow, then they may be able to develop a theory. In science, theories often develop slowly over many years and even over many generations of scientists.

Chapter 2

Solving Problems in the Chemistry of Life

You may be studying biology, but a little basic chemistry is essential to understanding life. Biology has to do with anything concerning living things, including their physical structure, which is where chemistry comes in. Chemistry has to do with the matter that makes up the physical world. In this chapter, you find out about all the building blocks, called *elements,* that comprise you and everything you see.

Mapping Atoms

Every living thing is made of matter, and all matter is made of elements. *Matter* is anything that takes up space and can be weighed. Different types of matter, called *elements,* have different properties. For example, you're probably familiar with the properties of the elements called metals.

An *atom* is the smallest whole, stable piece of an element that still has all the properties of that element. Every atom actually contains even smaller pieces known collectively as *subatomic particles.* These include protons, neutrons, and electrons (and even quarks, mesons, leptons, and neutrinos). You can't remove subatomic particles from an atom without destroying the atom.

Here's the basic breakdown of an atom's structure (you can see the visual in Figure 2-1; it's the creation of Danish scientist Niels Bohr):

✔ **The core of an atom, called the *nucleus,* contains two kinds of subatomic particles: protons and neutrons.** Both have a mass equal to 1 atomic mass unit (amu), but only one carries any kind of charge. *Protons* carry a positive charge, but *neutrons* have no charge (they're neutral). Because the protons are positive and the neutrons have no charge, the net charge of an atom's nucleus is positive.

✔ **Clouds of electrons surround the nucleus.** *Electrons* carry a negative charge but have essentially no mass. In an atom, the number of electrons equals the number of protons, balancing the electrical charge and making atoms neutral.

Atoms become *ions* when they gain or lose electrons; ions are essentially charged atoms. *Positive (+) ions* have more protons than electrons; *negative (−) ions* have more electrons than protons. Figure 2-1 shows how sodium and chlorine atoms become ions.

A. Bohr's model of an atom: Carbon used as an example.

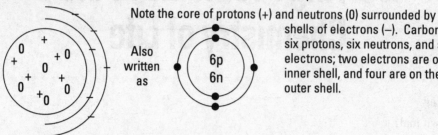

Note the core of protons (+) and neutrons (0) surrounded by shells of electrons (−). Carbon has six protons, six neutrons, and six electrons; two electrons are on the inner shell, and four are on the outer shell.

Also written as

6p
6n

B. Sodium and chloride ions joining to form table salt. The sodium ion has a positive charge because there's one more proton than electrons, so the overall charge is positive. The chloride ion is negative because after it accepts the electron from sodium, it then has one more electron than protons (18 versus 17), so the overall charge is negative. Together, though, NaCl is neutral because the "plus 1" charge is balanced by the "minus 1" charge.

Figure 2-1:
The Bohr model of an atom's structure. (B) An atom of sodium (sodium ion: Na⁺) joining an atom of chlorine (chloride ion: Cl⁻) to create the compound sodium chloride (table salt). (C) Two atoms of oxygen (O) combine to form one molecule of oxygen gas (O₂).

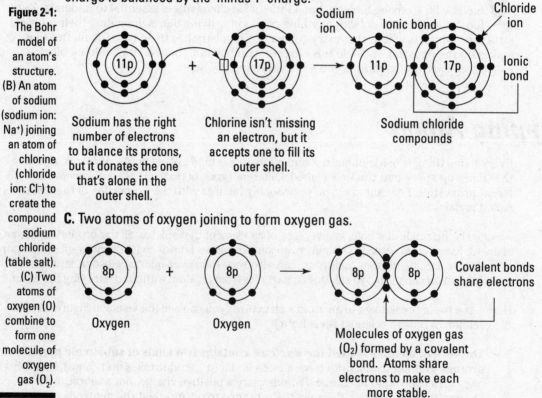

Sodium has the right number of electrons to balance its protons, but it donates the one that's alone in the outer shell.

Chlorine isn't missing an electron, but it accepts one to fill its outer shell.

Sodium ion
Ionic bond
Chloride ion

11p   17p

Ionic bond

Sodium chloride compounds

C. Two atoms of oxygen joining to form oxygen gas.

8p   +   8p   →   8p   8p

Oxygen   Oxygen

Covalent bonds share electrons

Molecules of oxygen gas (O₂) formed by a covalent bond. Atoms share electrons to make each more stable.

1. An atom has 62 protons. How many electrons would it have?

a. 62

b. 31

c. Can't tell from this information

2. A particle of matter has 19 protons, 19 neutrons, and 18 electrons. What is it?

a. An atom

b. A positively charged ion

c. A negatively charged ion

Elemental Thinking

An *element* is a substance that can't be broken down into something different by chemical means. Think of elements as pure chemical substances. All the known elements are organized into the *periodic table of elements* (shown in Figure 2-2), which is organized into periods and groups:

- **Each row of the table is called a *period*.** Moving across the table horizontally, you go from metals to nonmetals, with heavy metals in the middle.
- **Each column is called a family or *group*.** Elements within the same family/group have similar properties. The size of the atom increases from top to bottom within each column.

Each element in the table has a number, called the *atomic number,* that shows the number of protons in the nucleus of each atom of that element.

1 H Hydrogen 1.01																	2 He Helium 4.00
3 Li Lithium 6.94	4 Be Beryllium 9.01											5 B Boron 10.81	6 C Carbon 12.01	7 N Nitrogen 14.01	8 O Oxygen 16.00	9 F Fluorine 19.00	10 Ne Neon 20.18
11 Na Sodium 22.99	12 Mg Magnesium 24.31											13 Al Aluminum 26.98	14 Si Silicon 28.09	15 P Phosphorus 30.97	16 S Sulfur 32.06	17 Cl Chlorine 35.45	18 Ar Argon 39.95
19 K Potassium 39.10	20 Ca Calcium 40.08	21 Sc Scandium 44.96	22 Ti Titanium 47.90	23 V Vanadium 50.94	24 Cr Chromium 52.00	25 Mn Manganese 54.94	26 Fe Iron 55.85	27 Co Cobalt 58.93	28 Ni Nickel 58.71	29 Cu Copper 63.55	30 Zn Zinc 65.38	31 Ga Gallium 69.72	32 Ge Germanium 72.59	33 As Arsenic 74.92	34 Se Selenium 78.96	35 Br Bromine 79.90	36 Kr Krypton 83.80
37 Rb Rubidium 85.47	38 Sr Strontium 87.62	39 Y Yttrium 88.91	40 Zr Zirconium 91.22	41 Nb Niobium 92.91	42 Mo Molybdenum 95.94	43 Tc Technetium (99)	44 Ru Ruthenium 101.07	45 Rh Rhodium 102.91	46 Pd Palladium 106.42	47 Ag Silver 107.87	48 Cd Cadmium 112.41	49 In Indium 114.82	50 Sn Tin 118.69	51 Sb Antimony 121.75	52 Te Tellurium 127.60	53 I Iodine 126.90	54 Xe Xenon 130.30
55 Cs Cesium 132.91	56 Ba Barium 137.34	57 La Lanthanum 138.91	72 Hf Hafnium 178.49	73 Ta Tantalum 180.95	74 W Tungsten 183.85	75 Re Rhenium 186.21	76 Os Osmium 190.2	77 Ir Iridium 192.22	78 Pt Platinum 195.09	79 Au Gold 196.97	80 Hg Mercury 200.59	81 Tl Thallium 204.37	82 Pb Lead 207.19	83 Bi Bismuth 208.98	84 Po Polonium (210)	85 At Astatine (210)	86 Rn Radon (222)
87 Fr Francium (223)	88 Ra Radium (226)	89 Ac Actinium (227)	104 Rf Rutherfordium (257)	105 Db Dubnium (260)	106 Sg Seaborgium (263)	107 Bh Bohrium (262)	108 Hs Hassium (265)	109 Mt Meitnerium (266)									

Lanthanides	58 Ce Cerium 140.12	59 Pr Praseodymium 140.91	60 Nd Neodymium 144.24	61 Pm Promethium (147)	62 Sm Samarium 150.35	63 Eu Europium 151.96	64 Gd Gadolinium 157.25	65 Tb Terbium 158.93	66 Dy Dysprosium 162.50	67 Ho Holmium 164.93	68 Er Erbium 167.26	69 Tm Thulium 168.93	70 Yb Ytterbium 173.04	71 Lu Lutetium 174.97
Actinides	90 Th Thorium (232)	91 Pa Protactinium (231)	92 U Uranium (238)	93 Np Neptunium (237)	94 Pu Plutonium (242)	95 Am Americium (243)	96 Cm Curium (247)	97 Bk Berkelium (247)	98 Cf Californium (251)	99 Es Einsteinium (254)	100 Fm Fermium (257)	101 Md Mendelevium (258)	102 No Nobelium (259)	103 Lr Lawrencium (260)

Figure 2-2: The periodic table of elements.

All atoms of an element have the same number of protons, but the number of neutrons can change. If the number of neutrons is different between two atoms of the same element, the atoms are called *isotopes* of the element.

The total mass of one atom of an isotope is indicated by its *mass number,* which is often written as a superscript next to the atomic symbol.

Q. How many protons would be found in an atom of carbon-14 (^{14}C)?

A. Atoms of carbon-14 still have 6 protons (because all carbon atoms have 6 protons). Their mass is 14, so they must have 8 neutrons, because $14 - 6 = 8$.

The *atomic mass* of an element is the average mass of all the isotopes of that element, taking into account their relative abundance. If you look back at the periodic table in Figure 2-2, you see that the atomic mass of carbon (written underneath the letter C) is 12.01. This number tells you that if you took the average of the mass of all the carbon atoms on Earth, they'd average out to 12.01. The most stable isotope of carbon is carbon-12, so it's more abundant than carbon-14. (When you average the mass of lots of atoms of carbon-12 with some of carbon-14, you get a number slightly larger than 12.)

3. What's the atomic symbol for potassium?

   a. P

   b. K

   c. Pt

   d. Kr

4. How many protons does an atom of nitrogen have?

   a. 7

   b. 14

   c. It depends on which isotope of nitrogen it is.

5. What's the atomic mass of calcium?

   a. 20

   b. 40.08

   c. 6.02×10^{23}

6. Deuterium (^{2}H) is the isotope of hydrogen that's used to make heavy water for nuclear power plants. The mass number of deuterium is 2. How many neutrons does it have?

   a. 1

   b. 2

   c. 3

   d. 4

Figuring Out Molecules

When you start putting atoms together, you get more complex forms of matter, such as molecules and compounds. *Molecules* are made of two or more atoms, and *compounds* are molecules that contain at least two different elements. For example, oxygen gas is a molecule because its formula is O_2, indicating that it's made of two oxygen atoms joined together. It's not a compound, though, because both atoms are the same element.

Atoms join together by forming chemical bonds. Three types of bonds hold together the molecules that make up living things:

✔ **Ionic bonds** hold ions joined together by their opposite electrical charges. Ionic reactions occur when atoms combine and lose or gain electrons. Figure 2-1 shows how an ionic bond forms between the positively charged sodium ion (Na^+) and the negatively charged chloride ion (Cl^-).

✔ **Covalent bonds** form when atoms share electrons. Each shared pair of electrons is one covalent bond, so the two pairs of shared electrons in a molecule of oxygen gas have a *double bond* (refer to Figure 2-1). Covalent bonds are extremely important in biology because they hold together the backbones of all biological molecules. *Polar covalent bonds* form when two atoms share electrons, but not evenly. For example, in the polar covalent bonds in water molecules, the electrons spend more time near the nucleus of the oxygen atom and less time near the nuclei of the hydrogen atoms. This gives water molecules an unequal, or *polar*, distribution of charge. Figure 2-3 shows these slight negative and positive charges with the symbols δ^+ and δ^-.

✔ **Hydrogen bonds** are the weak electrical attractions that form between polar molecules or between polar groups of molecules. Individual hydrogen bonds are weak, but many of them together hold together important biological molecules like DNA and proteins.

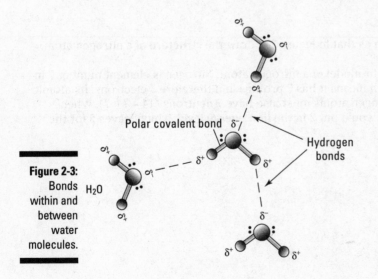

Figure 2-3:
Bonds
within and
between
water
molecules.

H_2O

Polar covalent bond

Hydrogen
bonds

The way that atoms interact or *react* with other atoms depends mainly on two factors:

- ✔ **The number of electrons in the atom.** Electrons arrange themselves in layers, called *electron shells* or *energy levels,* around the nucleus of the atom. Figure 2-1 shows these electron shells as the circles that surround the nucleus of each atom. The electrons fill the shells beginning with the energy level closest to the nucleus and then work their way out one level at a time. If the outermost energy level has a complete set of electrons, the atom becomes nonreactive or *inert.* All the elements in a row of the periodic table react with other atoms until they achieve the same stable arrangement of electrons in their energy levels as the inert gas at the far right in their row. Most of the elements needed in large amounts by living things are in the first three rows of the periodic table:

  - Elements in the first row, like hydrogen (H), only put electrons in the first energy level. This energy level is stable with two electrons.

  - Elements in the second row — like carbon (C), nitrogen (N), and oxygen (O) — put electrons in the first two energy levels. The second energy level is stable when it contains eight electrons.

  - Elements in the third row — like sodium, phosphorus, sulfur, and chlorine — put electrons in the first three energy levels. The third energy level is stable when it contains eight electrons.

- ✔ **The *electronegativity,* or attraction for electrons, of the atom.** Elements on the left in the periodic table are lower in electronegativity, while elements on the right are higher in electronegativity. If two atoms with approximately equal electronegativity interact with each other, they tend to pull on each other's electrons evenly and end up sharing them in covalent bonds. On the other hand, if a weakly electronegative atom interacts with a strongly electronegative atom, the stronger atom may just pull electrons away from the weaker one.

I know that's a lot of information to digest, so check out these examples to help you get the hang of how atoms interact.

Q. Using the same notation as that in Figure 2-1, draw the structure of a nitrogen atom.

A. Figure 2-4 shows the Bohr model of a nitrogen atom. Nitrogen is element number 7 in the periodic table, which means it has 7 protons and therefore 7 electrons. Its atomic mass is 14, so most nitrogen atoms must also have 7 neutrons (14 – 7 = 7). When arranging the electrons, you'd put 2 in the first energy level, which leaves 5 for the second energy level.

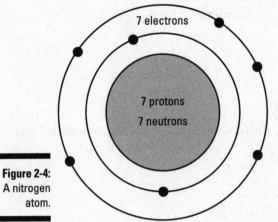

7 electrons

7 protons
7 neutrons

Figure 2-4:
A nitrogen
atom.

Q. Nitrogen reacts with hydrogen, joining with covalent bonds to form a compound with one nitrogen atom. How many hydrogen atoms would be in this compound?

A. From the preceding example, you know that nitrogen has 5 electrons in its outermost energy level. The second energy level is stable with 8 electrons, so each nitrogen atom accepts 3 electrons ($8 - 5 = 3$). Hydrogen is number 1 in the periodic table, so it has 1 proton and 1 electron. Therefore, 3 hydrogen atoms are needed for each nitrogen atom in order to form a stable compound. (The compound formed is NH_3, which is the formula for ammonia.)

Q. If lithium (Li) and chlorine (Cl) interact, will they form an ionic bond or a covalent bond? How many atoms of Li and Cl will react with each other?

A. Lithium is in the first column, or group, of the periodic table, so it's weakly electronegative, while chlorine is in the most electronegative group (the inert gases don't care about electrons because they're already stable). So when these two elements interact, chlorine will likely take electrons from lithium, just like it does from sodium (refer to Figure 2-1), and both lithium and chlorine will become ions that join with an ionic bond.

Lithium is atomic number 3, so it has 3 protons and 3 electrons. Two electrons fill its inner energy level, leaving just 1 in the second level. Chlorine is atomic number 17, so it has 17 protons and 17 electrons. Two electrons fill the inner energy level, 8 fill the second energy level, and 7 go into the third energy level ($17 - 2 - 8 = 7$). The third energy level is stable with 8 electrons, so chlorine is just looking for 1 electron, which is just what 1 lithium atom has to give. So, the 2 atoms interact in a one-to-one ratio.

7. How many electrons does sulfur have in its outermost energy level?

a. 2

b. 4

c. 6

d. 8

8. Carbon reacts with hydrogen, joining with covalent bonds to form a compound with 1 carbon atom. How many hydrogen atoms would be in this compound?

a. 2

b. 3

c. 4

d. 5

9. Calcium and chlorine atoms come into contact with each other and react to form a compound. Will this compound have ionic or covalent bonds?

a. Ionic

b. Covalent

10. Calcium and chlorine atoms come into contact with each other and react to form a compound. What is the chemical formula for this compound?

a. CaCl

b. Ca_2Cl

c. $CaCl_2$

d. Ca_3Cl_2

Recognizing Macromolecules

All living things rely pretty heavily on one particular element: carbon. The carbon atom is the central focus of *organic chemistry,* which is the chemistry of living things. In fact, scientists call the carbon-containing molecules that make up living things *organic molecules.* The carbon atoms in organic molecules form covalent bonds with other atoms like hydrogen, nitrogen, sulfur, and oxygen.

Carbon atoms join with other atoms to form four types of large molecules, called *macromolecules,* that form the structures of living things:

✔ **Carbohydrates** are important sources of energy and structural materials of living things.

✔ **Proteins** perform many important functions and are also important in structure and movement.

✔ **Nucleic acids** store and transmit information.

✔ **Lipids** store and provide energy, form structures, function in signaling, and can also help insulate organisms.

a Glucose

```
        H   O
         \ //
          C
          |
   H —— C —— OH
          |
  HO —— C —— H
          |
   H —— C —— OH
          |
   H —— C —— OH
          |
   H —— C —— OH
          |
          H
```

c Oligosaccharide   d

glucose
unit

h proteins

living things. They provide structure to cells,
ction possible. Proteins keep your metabolism
onses.

ide chains found in the proteins of life on Earth.
amino acid. Notice that, like carbohydrates,
oxygen. But amino acids also contain another
he same basic structure along their backbone,
e chemical group called a *side chain,* which is
simplest side chain is just a hydrogen atom,
ing from longer chains of carbon and hydrogen
tructures.

n of an amino acid, you can recognize them
a nitrogen in the amino group (NH_3), a central
group (COOH).

oup and a carboxyl group.
e of the 20 side-chain

id would be aspartic acid.
le bonds. Specific proteins
nnected together. The
code.

REMEMBER

Most animals, including people, can't
the glucose molecules. Cellulose, or
untouched, helping to maintain your

12. The backbone of carbohydrates code.
element?

a. Nitrogen

b. Sulfur

c. Oxygen

d. Phosphorus

13. Which carbohydrate passes thr
down?

a. Cellulose

b. Glucose

c. Glycogen

d. Starch

es break down little monosaccharides like
also molecules into your
e entire body so that your cells
fe. Your cells take the

ino acids to link together in a certain order,
. One or more polypeptide chains fold into a
hed *protein.* After the protein forms, it does a
the body.

t kinds of proteins and what they do:

e rate of chemical reactions.

tissues.

ound cells and around the body.

against bacteria and viruses.

pass them along to cells.

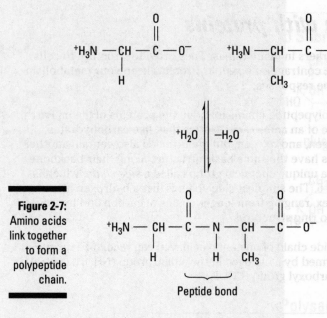

Figure 2-7:
Amino acids link together to form a polypeptide chain.

Peptide bond

14. Which element is found in proteins but not in most carbohydrates?

    a. Carbon

    b. Hydrogen

    c. Oxygen

    d. Nitrogen

15. What's the difference between a polypeptide and a protein?

16. Lactase is a protein that helps people break down the milk sugar lactose. What role does a protein like lactase perform for living things?

    a. Defense

    b. Transport

    c. Structure

    d. Metabolism

17. Collagen is a protein found in connective tissue, the tissue that joins muscles to bones. What role does a protein like collagen perform for living things?

    a. Defense

    b. Transport

    c. Structure

    d. Metabolism

Making plans with nucleic acids

Nucleic acids are specialists in information storage and retrieval. The information they contain is encoded in a pattern of four alternating molecules called nucleotides.

Nucleic acids are made up of strands of *nucleotides* (see Figure 2-8). Each nucleotide has three components of its own:

✔ A nitrogen-containing base called a *nitrogenous base*

✔ A sugar that contains five carbon molecules

✔ A phosphate group

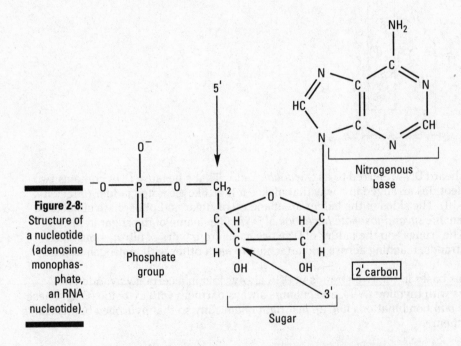

Figure 2-8:
Structure of a nucleotide (adenosine monophosphate, an RNA nucleotide).

Nucleotides are different from one another based on the nitrogenous base they contain. Consider DNA (short for *deoxyribonucleic acid*), the most famous nucleic acid. DNA stores the instructions for the structure and function of all living things, from bacteria to mushrooms to people. The nitrogenous bases that form DNA contain the nitrogenous bases adenine (A), guanine (G), cytosine (C), and thymine (T). Nucleotides containing these bases link together to form polynucleotide chains, as shown in Figure 2-9. The order of the chemical letters in the nucleotides spells out your genetic code. Your cells read your genetic code to get instructions on how to build your molecules (more on this in Chapter 7).

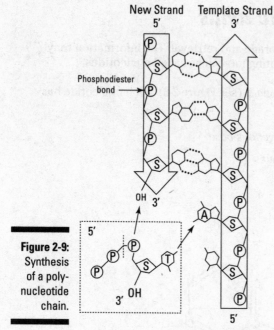

New Strand   Template Strand

Phosphodiester bond →

Figure 2-9:
Synthesis of a polynucleotide chain.

Illustration by Kathryn Born, M.A.

You may have heard DNA referred to as the *double helix*. That's because DNA contains two strands of nucleotides arranged in a way that makes it look like a twisted ladder (check it out in Figure 2-10). The sides of the ladder are made up of sugar and phosphate molecules, hence, the nickname *sugar-phosphate backbone* of DNA. (The name of the sugar in DNA is *deoxyribose*.) The "rungs" on the ladder of DNA are made from pairs of nitrogenous bases from the two strands, reaching across and attaching to each other with hydrogen bonds.

The nitrogenous bases in the two strands of DNA always join in a certain way: Adenine always partners with thymine (A-T), and guanine always partners with cytosine (G-C). These particular *base pair* combinations line up just right chemically so that hydrogen bonds can form between them.

Another important nucleic acid in cells is called RNA, short for *ribonucleic acid.* RNA molecules play an important role in the creation of new proteins (see Chapter 7).

The structure of RNA is slightly different from that of DNA:

- ✔ RNA molecules have only one strand of nucleotides.
- ✔ The nitrogenous bases used are adenine (A), guanine (G), cytosine (C), and uracil (U) (rather than thymine).
- ✔ The sugar in RNA is ribose (not deoxyribose).

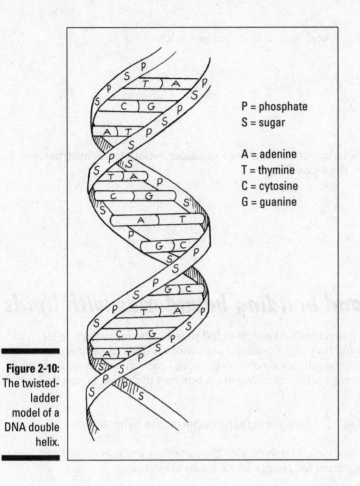

P = phosphate
S = sugar

A = adenine
T = thymine
C = cytosine
G = guanine

Figure 2-10:
The twisted-
ladder
model of a
DNA double
helix.

18. You isolate the DNA from some cells and chemically analyze it, finding that its
nitrogenous bases are 14% adenine. What percentage of the nitrogenous bases would
be cytosine?

a. 14%

b. 28%

c. 36%

d. 72%

19. Which nitrogenous base is found only in DNA and not in RNA?

   a. Adenine (A)

   b. Thymine (T)

   c. Cytosine (C)

   d. Guanine (G)

20. If you think of the double helix of DNA as a twisted ladder, what chemical structures make up the handrails of the ladder?

   a. Sugars only

   b. Phosphates only

   c. Nitrogenous bases only

   d. Sugars and phosphates

Storing energy and building boundaries with lipids

Lipids are a diverse group of molecules that are grouped together because they're all *hydrophobic* molecules, meaning they don't mix well with water. Lipids are sometimes called *hydrocarbons* because they're primarily made of hydrogen and carbon atoms, with small amounts of oxygen and other atoms. Bonds between carbon and hydrogen atoms are very rich in energy.

Here are a few examples of lipids that are particularly important to living things:

- **Triglycerides:** Fats and oils, called triglycerides because they contain three fatty acids (see Figure 2-11), are important for energy storage and insulation.

- **Phospholipids:** These lipids have an important structural function for living things because they're part of the membranes of cells (see Chapter 3 for more on cell membranes). They look similar to triglycerides, but instead of one of the fatty acid chains, they have a *hydrophilic* head group that mixes easily with water.

- **Sterols:** These lipid compounds, consisting of four connecting carbon rings and a functional group that determines the steroid (see Figure 2-12 for an example), generally create hormones.

Whether a triglyceride is a fat or an oil depends on the bonds between the carbon and hydrogen atoms.

- Fats contain lots of single covalent bonds between their carbon atoms. These *saturated* bonds pack tightly, so fats are solid at room temperature.

- Oils contain lots of double covalent bonds between their carbon atoms. These *unsaturated* bonds don't pack tightly, so oils are liquid at room temperature.

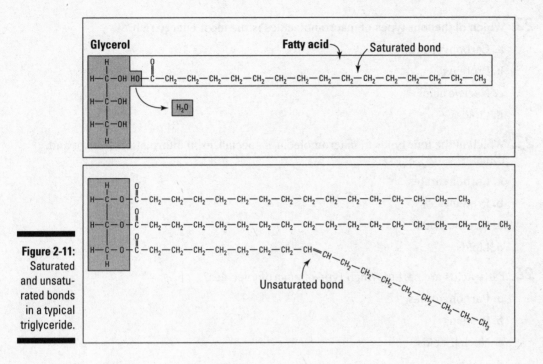

Figure 2-11: Saturated and unsaturated bonds in a typical triglyceride.

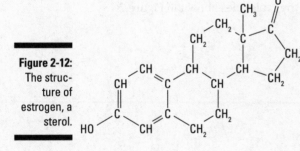

Figure 2-12: The structure of estrogen, a sterol.

21. Plants that live in environments that get cold in winter have to be able to cope with the changes in temperature. One problem they encounter is that the cold temperatures tend to make their lipids get thick and more solid. In order to survive, the plants must have a way to keep the phospholipids in their plasma membranes from forming a solid layer. Which of the following two strategies would help plants keep their plasma membranes fluid even when temperatures are falling?

a. Build phospholipids with lots of unsaturated bonds in their fatty acids.

b. Build phospholipids with lots of saturated bonds in their fatty acids.

22. Which of the four types of macromolecules is the most energy-rich?

   a. Carbohydrates

   b. Proteins

   c. Nucleic acids

   d. Lipids

23. Which of the four types of macromolecules specializes in information storage and transfer?

   a. Carbohydrates

   b. Proteins

   c. Nucleic acids

   d. Lipids

24. Fatty acids are part of which type of macromolecules?

   a. Carbohydrates

   b. Proteins

   c. Nucleic acids

   d. Lipids

25.–32. Use the terms that follow to identify the molecules shown in Figure 2-13.

   a. Carbohydrate (oligosaccharide)

   b. Amino acid

   c. Nucleotide (from DNA)

   d. Lipid (fatty acid)

   e. Lipid (sterol)

   f. Polynucleotide chain

   g. Lipid (phospholipid)

   h. Carbohydrate (monosaccharide)

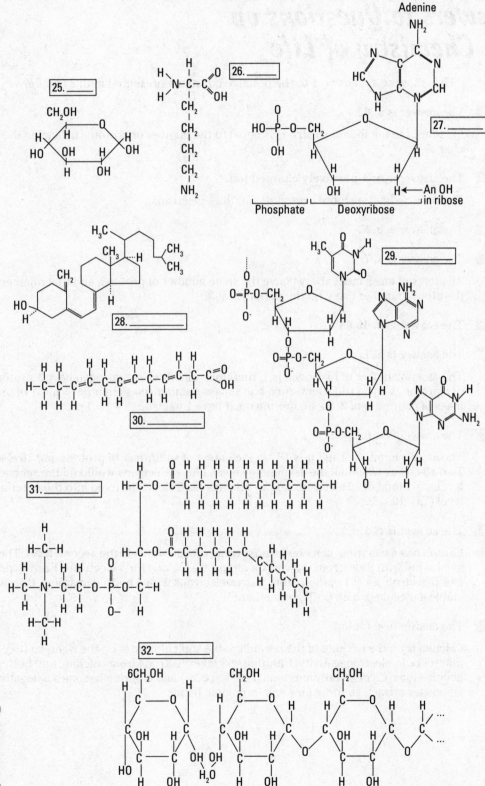

Figure 2-13:
Identifying
macro-
molecules.

Answers to Questions on the Chemistry of Life

The following are answers to the practice questions presented in this chapter.

1 The answer is **a. 62.**

In atoms, the number of electrons is equal to the number of protons, balancing the electrical charge.

2 The answer is **b. A positively charged ion.**

The particle has one more proton than it does electrons.

3 The answer is **b. K.**

4 The answer is **a. 7.**

All atoms of an element always have the same number of protons, and that number is equal to the atomic number listed on the periodic table.

5 The answer is **b. 40.08.**

6 The answer is **a. 1.**

The atomic number of hydrogen is 1; therefore, all atoms of hydrogen have 1 proton. The mass of a proton is equal to 1. Deuterium has a mass number (mass number = mass of protons + mass of neutrons) of 2. So, deuterium must have 1 neutron (2 – 1 = 1).

7 The answer is **c. 6.**

The atomic number of sulfur is 16, so each atom of sulfur has 16 protons and 16 electrons. Two electrons would fill the first energy level, plus 8 electrons would fill the second energy level, for a total of 10 electrons so far. That leaves 6 electrons to go into the outermost energy level (16 – 10 = 6).

8 The answer is **c. 4.**

Carbon has 4 electrons in its outermost energy level, which is the second level. The second level is full with 8 electrons, so carbon atoms are looking for 4 electrons. Each hydrogen atom has 1 electron. So, if 1 carbon atom shares electrons with 4 hydrogen atoms, they will form a stable molecule (which is CH_4 or methane).

9 The answer is **a. Ionic.**

Calcium is on the left side of the periodic table and chlorine is on the right, so they show a big difference in electronegativity. Chlorine will take electrons from calcium, and both atoms will become ions. Calcium becomes a positive ion (Ca^+) and chlorine becomes a negative ion (Cl^-). Opposites attract, so these ions will form ionic bonds.

10 The answer is **c. $CaCl_2$**.

Each calcium atom has 2 electrons in its outermost energy level, which is the second level. So calcium needs to either give up those 2 electrons or get 6 electrons in order to be inert. Chlorine has 7 electrons in its outermost energy level, so it only needs to get 1 to become inert. Because of the big difference in electronegativity between these atoms, chlorine will take electrons from calcium. But each chlorine atom can only take 1 electron into its outermost energy level. Thus, it will take 2 chlorine atoms to accept the available electrons from 1 calcium atom.

11 The answer is **b. Proteins.**

12 The answer is **c. Oxygen.**

13 The answer is **a. Cellulose.**

14 The answer is **d. Nitrogen.**

15 Polypeptides are chains of amino acids. Proteins are folded up chains of amino acids that are ready to do a job for a cell. In other words, proteins are the actual molecules that do the jobs. Some proteins are made of only one folded polypeptide chain, but some proteins won't work unless two or more polypeptide chains fold up together.

16 The answer is **d. Metabolism.**

17 The answer is **c. Structure.**

18 The answer is **c. 36%.**

If A is 14%, then T is also 14% (A always pairs with T), for a total A-T percentage of 28%. $100 - 28 = 72\%$ remaining for the C-G percentage. Half of 72 is 36.

19 The answer is **b. Thymine (T).**

RNA contains the base Uracil (U) instead of thymine.

20 The answer is **d. Sugars and phosphates.**

21 The answer is **a. Build phospholipids with lots of unsaturated bonds in their fatty acids.**

Unsaturated fatty acids don't pack tightly and so remain more fluid. Cold temperatures cause lipids to thicken, so a cell would want to make fluid plasma membranes in order to be cold-resistant.

22 The answer is **d. Lipids.**

23 The answer is **c. Nucleic acids.**

24 The answer is **d. Lipids.**

25 – 32 The following is how Figure 2-13 should be labeled:

25 **h. Carbohydrate (monosaccharide);** 26 **b. Amino acid;** 27 **c. Nucleotide (from DNA);** 28 **e. Lipid (sterol);** 29 **f. Polynucleotide chain;** 30 **d. Lipid (fatty acid);** 31 **g. Lipid (phospholipid);** 32 **a. Carbohydrate (oligosaccharide).**

Chapter 3

Identifying Cell Parts and Understanding Their Functions

In This Chapter

▶ Looking at a cell's main parts and functions

▶ Breaking down animal cells

▶ Seeing how plant cells differ from animal cells

▶ Focusing on prokaryotes

*E*very living thing — from tiny bacteria to big blue whales — is made of cells. Most bacteria are made of just one cell, while your body has approximately 10 trillion cells, but both you and bacteria share the same fundamental characteristics of life. Each characteristic of life — like getting energy, moving, and reproducing — depends on a specific structure in the cell. In this chapter, you explore the parts of cells and their functions.

Introducing Cells

Cells are sacs of thick fluid that are reinforced by proteins and surrounded by a flexible barrier called a *plasma membrane.* Scientists call the thick fluid inside cells the *cytoplasm.* Inside the cytoplasm float chemicals and *organelles,* structures inside cells that are used during metabolic processes. Some organelles have their own membrane boundaries, while others don't. The *nucleus* is a particularly large and distinctive membrane-bound organelle found in your cells and the cells of other animals (like the one in Figure 3-1), plants, fungi, algae, and some single-celled organisms. The nucleus is a special compartment that contains the cell's genetic material (DNA).

A cell is the smallest part of an organism that retains characteristics of the entire organism. For example, a cell can take in fuel, get energy, and eliminate waste, just like the organism as a whole. Because cells can perform all the functions of life, the cell is the smallest unit of life.

Scientists categorize cells in different ways depending on whether they're interested in genetic relationships, structure, or function. Based on fundamental cellular organization, scientists recognize two main categories of cells:

    ✔ **Eukaryotes** have membrane-bound organelles and a nucleus that houses their genetic material (DNA). Plants, animals, algae, and fungi are all eukaryotes. (Refer to the drawing of an animal cell in Figure 3-1.)

✔ **Prokaryotes** don't have membrane-bound organelles or a nucleus; their DNA sits right in the cytoplasm. Bacteria and archaea are prokaryotes. Compare the drawing of a bacterial cell in Figure 3-2 with the drawing of an animal cell in Figure 3-1. Flip to the section "Peeking at Prokaryotes" later in the chapter to find out more about them.

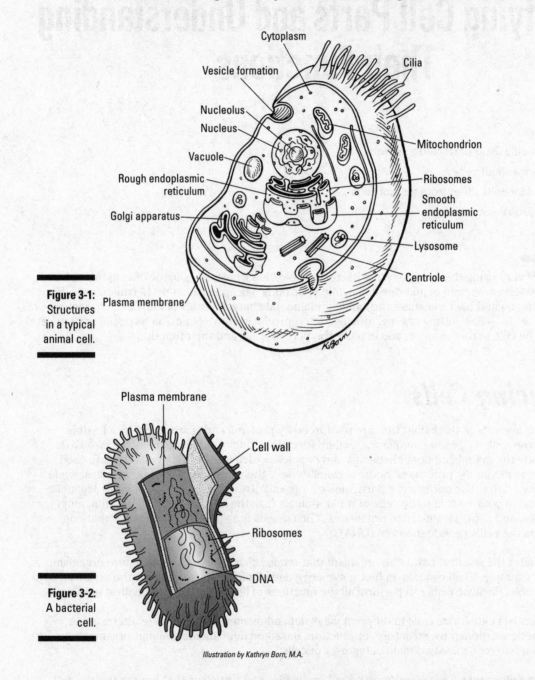

Cytoplasm

Vesicle formation

Cilia

Nucleolus

Nucleus

Mitochondrion

Vacuole

Rough endoplasmic reticulum

Ribosomes

Smooth endoplasmic reticulum

Golgi apparatus

Lysosome

Centriole

Figure 3-1: Structures in a typical animal cell.

Plasma membrane

Plasma membrane

Cell wall

Ribosomes

DNA

Figure 3-2: A bacterial cell.

Illustration by Kathryn Born, M.A.

Another way to tell the difference between eukaryotes and prokaryotes is cell size. Eukaryotic cells are typically about 10 times larger than prokaryotic cells (although not always, so be sure to look for other key features like whether the cell has a nucleus and membrane-bound organelles!). Eukaryotic cells are usually between 10 and 100 micrometers (μm) wide, while prokaryotes are between 1 and 10 μm in size. A micrometer is 1/1000 of a millimeter, and a millimeter is about as big as a period on this page. If you think about those sizes, you'll understand pretty quickly why we need microscopes in order to see cells!

1. A scientist discovers a brand-new type of single-celled organism growing on the surface of a fish from the deep ocean. The cell is 1000 μm (1 millimeter) in size. Is the cell a prokaryote or a eukaryote?

   a. It's a prokaryote because it's single-celled.

   b. It's a eukaryote because it's so big.

   c. I can't tell from this information.

Holding it all together: The plasma membrane

The plasma membrane that separates and defines cells is pretty much the same no matter what type of cell you're looking at. The job of the plasma membrane is to separate the chemical reactions that occur inside cells from the chemicals outside the cell and to control what enters and leaves cells. Scientists say that the plasma membrane is *selectively permeable:* selective because it chooses what can go in and out and permeable because it can be crossed.

Thinking of the plasma membrane as an international border where customs agents control what enters and leaves a particular country is a good way of remembering the plasma membrane's function.

Plasma membranes are made of several different components, much like a mosaic work of art. Because membranes are a mosaic of different types of molecules, and because they're flexible and fluid, scientists call the description of membrane structure the *fluid-mosaic model.* Check out the drawing in Figure 3-3 to help you visualize all the molecules that make up a plasma membrane:

   ✔ **The phospholipid bilayer forms the foundation of the plasma membrane.** *Phospholipids* are a special kind of lipid that have *hydrophilic heads* that are attracted to water and *hydrophobic tails* that repel water. Two layers of phospholipids stack on top of each other to form the bilayer. Because cells reside in a watery solution and contain a watery solution inside of them (cytoplasm), the plasma membrane forms a sphere around each cell so that the water-attracting heads are in contact with the fluids inside and outside of the cell, while the water-repelling tails are sandwiched on the inside of the membrane (see Figure 3-3).

   ✔ **Proteins are the second major component of plasma membranes.** The proteins are embedded in the phospholipid bilayer, but they can drift in the membrane like ships sailing through an oily ocean. Many plasma membrane proteins are large enough to cross the entire bilayer, so they're in contact with both the outside and inside environment of the cell. These proteins may help materials cross the plasma membrane, or they may act like little satellite dishes, picking up signals outside the cell and passing the information inside.

   ✔ **Cholesterol is a minor component of the membrane.** Cholesterol sits between the phospholipids, where it helps stabilize membranes and prevents them from freezing when the temperature is low. Like all sterols, cholesterol molecules consist of four joined rings.

   ✔ **Small amounts of carbohydrates attach to the outer surfaces of plasma membranes.** These short carbohydrate chains are important cellular markers, determining characteristics such as your blood type.

2.–6. Use the following terms to label the diagram of the plasma membrane in Figure 3-3.

    a. Hydrophilic head

    b. Hydrophobic tail

    c. Protein

    d. Carbohydrate

    e. Cholesterol

    f. Phospholipid bilayer

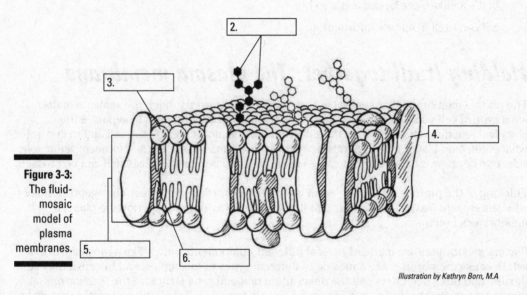

Figure 3-3: The fluid-mosaic model of plasma membranes.

Illustration by Kathryn Born, M.A

7. Imagine that the membrane in Figure 3-3 is just a small piece of a sphere of membrane that surrounds and defines a cell. Which side of the membrane would be pointed to the environment outside of the cell, the side toward the top of the figure or the side toward the bottom? How can you tell?

Getting in and out of cells

Cells are busy places. They manufacture materials and produce wastes that need to be shipped out, and they need to take up materials such as food and signals. These important exchanges take place at the plasma membrane. Small, hydrophobic molecules (like oxygen and carbon dioxide) can easily scoot through the hydrophobic tails of the phospholipid bilayer. Hydrophilic molecules (like ions) and larger molecules (think food and hormones) need the aid of *transport proteins,* proteins that span the membrane and create passageways for materials to cross the phospholipid bilayer.

Materials can pass through the plasma membrane either passively or actively. *Passive transport* doesn't require energy input from the cell, whereas *active transport* does. The two methods of passive transport are

✔ **Diffusion:** The passive transport of molecules other than water from an area where they're highly concentrated to an area where they're less concentrated. To go from a high concentration to a low concentration, the molecules need only spread themselves, or *diffuse,* across the membrane, separating the areas of concentration.

✔ **Osmosis:** The diffusion of water across a membrane. Osmosis works the same way as diffusion, but it can be a little confusing because the movement of water is affected by the concentration of substances, called *solutes,* that are dissolved in the water. Basically, water moves from areas where it's more concentrated, or more pure, to areas where it's less concentrated or pure and has more solutes in it.

Try thinking about osmosis in terms of the solutes: Water moves toward the area with the greatest concentration of solutes. You can remember this with the phrase "salt sucks." The fluid with the highest solute (salt) concentration will suck water towards it.

Contrary to passive transport, active transport moves molecules from areas where they're less concentrated to areas where they're more concentrated. Active transport enables cells to use their stored energy to concentrate molecules inside or outside of the cell.

You can think of passive and active transport like rolling a ball on a hill. Imagine that the hill is made of molecules; the low part of the hill has few molecules, while the big part of the hill has lots of stacked molecules. If you want to roll a ball downhill (or let molecules flow from where they're piled up to where they're not), you don't have to use any energy. If you want to roll the ball uphill (or push molecules from where they're in low concentration to where they're in high concentration), you have to put in some energy.

8.-10. Use the terms that follow to identify the type of transport occurring in the cell. If more than one answer is possible, choose the most specific answer.

    a. Passive transport

    b. Osmosis

    c. Active transport

8. A cell that contains 1.0% glucose (a solute) and is surrounded by a solution that's 0.5% glucose is moving glucose into the cell.

9. A cell containing 0.1% sodium chloride (NaCl) is placed in seawater that contains 3.0% NaCl, causing water to leave the cell.

10. A cell containing 0.1% sodium chloride (NaCl) is placed in seawater that contains 3.0% NaCl, causing sodium ions to enter the cell.

11. A cell is using active transport to pump chloride ions outside the cell. If this is the only thing the cell is doing, what would be happening to the amount of energy stored in the cell?

    a. It would increase.

    b. It would decrease.

    c. It would stay the same.

12. Your blood consists of blood cells and solutes floating in a solution called *plasma.* The concentration of solutes in blood plasma is about 0.9%. If a person undergoing medical treatment needs fluids, he may be given an intravenous (IV) injection of 0.9% saline (if you've never experienced this, you've probably seen it on TV). Why is it important for the concentration of solutes in IV saline to be the same as the concentration of solutes in blood plasma? What would happen to a person's red blood cells if the person were given an injection of pure water (0% saline)? What would happen if the person were given a higher concentration of solutes (5% saline)?

13. Dialysis tubing is used to filter the body fluids of people who suffer from kidney disease. The tubing contains many tiny holes that allow passive transport of molecules small enough to pass through the holes. A student places a solution of 10% starch inside a length of dialysis tubing and seals off the ends. She places the starch-containing solution into a glass container of pure water. Starch molecules are too big to fit through the holes in dialysis tubing. If the student leaves the tubing in the water overnight, which of the following will occur?

a. Starch will move out of the tubing and into the water, while water moves into the tubing by osmosis.

b. Only water will move into the tubing by osmosis.

c. Only starch will move out of the tubing and into the water by passive transport.

d. Neither starch nor water will move.

Creating proteins: Ribosomes

All cells have *ribosomes,* small structures in the cytoplasm that act as workbenches for the construction of proteins. The instructions for proteins are copied from the DNA into a new molecule, called *messenger RNA* or *mRNA.* The mRNA leaves the nucleus and carries the instructions to the ribosomes out in the cytoplasm. The ribosomes then organize the mRNA and other molecules that are needed to help put proteins together (for the full scoop on how proteins are made, flip to Chapter 7).

14. What is the function of ribosomes?

a. To store information

b. To read DNA

c. To build mRNA

d. To make proteins

Taking a Tour of Animal Cells

In some ways, you're very different from the animals you see around you like dogs, birds, spiders, and worms. But on the cellular level, you and other animals are very much alike. Your bodies are made of organs, which are made of tissues, which are made of cells. And, just like you have organs that perform specific functions for your body, your cells have organelles that perform specific functions for the cell (look back at Figure 3-1 to see a sketch of the following cell parts):

- A *nucleus* that stores genetic information.

- A *plasma membrane* that encloses the cell and separates it from its environment, controlling what enters and exits the cell.

- *Ribosomes* in the cytoplasm that make proteins for use inside the cell.

✔ Internal membranes, such as the *endoplasmic reticulum* (ER), that manufacture proteins and lipids for the cell.

- Ribosomes attach to part of the ER, making it appear dotted, or rough. This *rough endoplasmic reticulum* (RER) specializes in making proteins that either become part of membranes or are shipped out of the cell.

- Scientists call the part of the ER that doesn't have ribosomes the *smooth endoplasmic reticulum* (SER). It specializes in making lipids for the cell and performs specialized tasks in different tissues.

✔ The *Golgi apparatus,* a stack of folded membranes that makes changes to the lipids and proteins made by the ER. The Golgi basically puts small chemical tags on these molecules, directing them to be shipped to particular locations within the cell.

✔ *Transport vesicles,* small spheres of membrane the cell uses to ship materials where they need to go. For example, a protein made by the RER may be placed inside a transport vesicle and shipped to the Golgi, where it would be modified and tagged for shipment elsewhere.

✔ *Lysosomes,* special spheres of membrane that contain digestive enzymes. When cells want to break down and recycle old parts or kill something like a bacterium, they make use of lysosomes.

✔ A *cytoskeleton,* made of long, tubular proteins, that has multiple functions:

- The proteins of the cytoskeleton act like scaffolding to reinforce and strengthen cells.

- Cytoskeletal proteins help cells move because they're located inside cellular projections called *flagella* and *cilia.* Flagella and cilia are hairlike structures that project from the surfaces of cells. When the cytoskeletal proteins inside these structures bend, they make flagella and cilia beat like little whips, making some cells (like human sperm) swim and other cells (like in your upper airway) push fluids along.

- Cells also use the cytoskeleton to circulate materials within the cell. For example, a cell may put proteins inside a vesicle and then send the vesicle traveling along cytoskeletal proteins like a boxcar on a train track. *Centrioles* are part of the cytoskeleton that associate with the structure used to move chromosomes around when cells reproduce.

✔ *Mitochondria,* organelles that combine oxygen and food to transfer the energy from food to a form that cells can use (stored in the energy carrier, ATP; flip to Chapter 4 for the details). Scientists call the process that transfers energy from food to a usable form for the cell *cellular respiration.*

When you're learning about cells for the first time, remembering all the cell parts and their functions can be hard. One really great way to help yourself remember something is to imagine something familiar and then link the familiar object to the name of the thing you're trying to memorize. For the parts of cells, one approach that works well is to imagine a cell as a city. Then think about each organelle in the cell and what it does in relation to something familiar in that city.

15.–23. Use the terms that follow to identify the organelle that matches each example or scenario. Some terms may be used more than once.

a. Nucleus

b. Mitochondria

c. Cytoskeleton

d. Rough endoplasmic reticulum

e. Smooth endoplasmic reticulum

f. Golgi apparatus

g. Lysosome

h. Transport vesicle

15. A structure that supports and strengthens the cell.

16. An animal cell has a defective organelle, such that it gets very little energy from food molecules.

17. A student looks through a microscope at this organelle and sees chromosomes that contain genetic material.

18. A cell breaks down a captured bacterium inside this organelle.

19. At this structure, a white blood cell makes a protein that it will release into your blood.

20. A cell uses this structure to wave little projections, pushing mucus through your digestive tract.

21. This organelle makes phospholipids for the plasma membrane.

22. A protein is traveling through the cell inside this sphere-shaped structure.

23. A lipid arrives at this organelle and is tagged with a chemical group, then sent to thenuclear membrane.

24. One of the cells in your pancreas is making the protein insulin. After the cell makes the protein, it ships the protein outside the cell to your bloodstream. Name all the structures that would be involved in this process, in order, beginning with the moment the blueprint for the protein (in mRNA) leaves the nucleus.

Checking Out Plant Cells

You probably think you're pretty different from a plant, and on the surface, you'd be right. But at the cellular level, you'd be surprised to find out how much you and your favorite plants have in common. Plant cells have almost everything that animal cells have, and they perform all the same functions as animal cells.

What makes plants seem so different to us is that they do just a little bit more. Whereas we animals have to eat to get food for energy and matter, plants can make their own food (using a special organelle called a *chloroplast;* read on to find out more).

Plant cells have three features not found in animal cells (compare Figure 3-4 with Figure 3-1 to see the differences):

✔ *Chloroplasts* are green organelles that use energy from sunlight, plus water and carbon dioxide, to make food. Scientists call the food-making process that occurs in the chloroplasts *photosynthesis.*

✔ A rigid *cell wall* outside of their plasma membrane. The cell wall helps join plant cells into tissues and helps give plants internal strength. The cell walls of woody plants like trees become hard so that woody plants can grow tall. In fact, the cell walls of dead plant cells are all that remain in the familiar wood in the buildings and furniture around you.

✔ A very large, central *vacuole,* which is basically a big membrane-enclosed water balloon. Plants store a variety of materials in their vacuoles, including wastes, colored pigments, and chemicals that discourage grazing animals like insects.

One of the most common mistakes people make when learning about plant cells is to think that they have chloroplasts *instead of* mitochondria. True, chloroplasts let plants capture energy from the sun and matter from the environment (carbon dioxide from the air, water from the soil) and store the matter and energy as food. But the chloroplasts can't break down this food. Mitochondria help cells break down food molecules so they can use the stored energy and matter. So plants need chloroplasts *and* mitochondria. You can think of chloroplasts as the organelle that makes the lunch and the mitochondrion as the organelle you need to eat the lunch (on a cellular level).

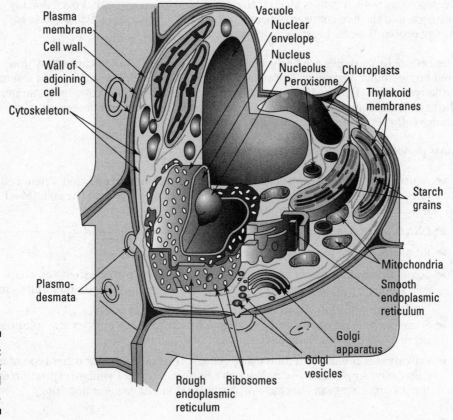

Figure 3-4: Structures in a typical plant cell.

25. Which structure is found in plant cells but not in animal cells?

　a. Ribosome

　b. Mitochondrion

　c. Cell wall

　d. Nucleus

26. Indian Pipe is a very unusual plant. Its leaves and stems are completely white; in fact, there's not a speck of green on its entire body. This plant can only grow near certain trees, from which it steals food. If you looked at the cells of Indian Pipe under a microscope, which of the following organelles would you expect to see?

   a. Chloroplasts but no mitochondria

   b. Mitochondria but no chloroplasts

   c. Chloroplasts and mitochondria

   d. Neither mitochondria nor chloroplasts

Peeking at Prokaryotes

Prokaryotes include cells you've probably heard of, such as the bacteria *E. coli* and *Streptococcus* (which causes strep throat), the blue-green algae that occasionally cause lake closures, and the live cultures of bacteria in yogurt. Prokaryotes also include some cells you've probably never heard of, such as *archaeans*.

The cells of prokaryotes are fairly simple in terms of structure because they don't have membrane-bound organelles like eukaryotic cells. (Refer to Figure 3-2 for an example of a prokaryotic cell.) Despite this difference in organization, prokaryotic cells can perform all the same basic functions as eukaryotic cells. Green bacteria and blue-green algae even do photosynthesis — they just don't do it in a chloroplast!

Most prokaryotes share these characteristics:

   ✔ A plasma membrane forms a selective barrier around the cell, and a rigid cell wall outside the plasma membrane provides additional support to the cell. (Most bacteria have cell walls, but some archaeans don't.)

   ✔ DNA is located in the cytoplasm, in an area called the *nucleoid*.

   ✔ Ribosomes make proteins in the cytoplasm.

   ✔ Some prokaryotes have cytoskeletal proteins that help strengthen and shape the cell. And some prokaryotes swim using flagella, but their flagella are built differently from eukaryotic flagella.

   ✔ Some bacteria do photosynthesis to make food. In these prokaryotes, photosynthesis occurs at the plasma membrane instead of in a chloroplast.

   ✔ Prokaryotes break down food using cellular respiration and another type of metabolism called *fermentation,* which doesn't require oxygen. Cellular respiration occurs at the plasma membrane because prokaryotes don't have mitochondria.

27. True or false: Prokaryotic cells don't have a nucleus, so they don't have any DNA.

28. You use a microscope to compare the cells of a green, photosynthetic bacterium and a green plant. Which structure would you expect to find in both cells?

   a. Chloroplast

   b. Cell wall

   c. Nucleus

   d. Mitochondrion

Answers to Questions on Cells

The following are answers to the practice questions presented in this chapter.

 The answer is **c. I can't tell from this information.**

Although eukaryotes are usually much larger than prokaryotes, you can't rely on size 100 percent. Some very small eukaryotes and some very large prokaryotes have both been discovered. Likewise, although most prokaryotes are single-celled, many eukaryotes are too. For example, many plankton in the ocean and freshwater habitats (and even the one on *SpongeBob SquarePants*) are single-celled eukaryotes. Also, a couple of very unusual prokaryotes group together at certain times in their lives and form a multicellular structure.

2 – **6** The following is how Figure 3-3 should be labeled:

2 **d. Carbohydrate;** 3 **a. Hydrophilic head;** 4 **c. Protein;** 5 **f. Phospholipid bilayer;** 6 **b. Hydrophobic tail**

7 The answer is **the side toward the top of the figure.**

You can tell because that's the side with attached carbohydrates. Carbohydrates only attach to the plasma membrane on the outside of the cell.

8 The answer is **c. Active transport.**

9 The answer is **b. Osmosis** (which is also a type of passive transport).

10 The answer is **a. Passive transport.**

11 The answer is **b. It would decrease.**

In order to do active transport, cells must supply energy. So the cell would have to use some of its stored energy to move the ions.

12 It's important to make sure any fluid injected into the body has the same solute concentration as the cells in the body so that the cells aren't damaged by the movement of water via osmosis. If a medical professional gave a person an injection of pure water, the solute concentration in the blood cells would be higher than that of the water. Water would move into the blood cells, causing them to swell up and then burst. If a medical professional gave a person an injection of saline that had a higher solute concentration than the blood cells, water would leave the cells and move into the plasma, causing the blood cells to shrivel.

13 The answer is **b. Only water will move into the tubing by osmosis.**

The concentration of solutes inside the tubing is greater than the concentration of solutes in the glass container, so water will move by osmosis into the tubing. Although the starch concentration inside the tubing is higher than that outside the tubing, the starch is too big to fit through the holes in the tubing, so the starch will be unable to move by passive transport.

14 The answer is **d. To make proteins.**

DNA stores the information for how to build the protein in the nucleus. That information is read and copied into mRNA inside the nucleus away from the ribosomes. Only the mRNA leaves the nucleus and travels to the ribosome.

15 The answer is **c. Cytoskeleton.**

16 The answer is **b. Mitochondria.**

17 The answer is **a. Nucleus.**

18 The answer is **g. Lysosome.**

19 The answer is **d. Rough endoplasmic reticulum.**

20 The answer is **c. Cytoskeleton.**

21 The answer is **e. Smooth endoplasmic reticulum.**

22 The answer is **h. Transport vesicle.**

23 The answer is **f. Golgi apparatus.**

24 The mRNA would travel to a ribosome, where protein synthesis would begin. The ribosome would attach to the rough ER and finish making the protein. The rough ER would wrap the protein in a vesicle, which would travel along the cytoskeleton to the Golgi. At the Golgi the protein would be chemically modified before being shipped off again in another vesicle. The vesicle would travel on the cytoskeleton to the plasma membrane. The vesicle would join with the plasma membrane to release the protein from the cell.

25 The answer is **c. Cell wall.**

Both plant and animal cells have all three other structures.

26 The answer is **b. Mitochondria but no chloroplasts.**

The plants are white and they can only grow when they can steal food, suggesting that they don't have any green chloroplasts for making their own food. However, because the plants can grow, they must have a way to get energy and matter from the food they steal, which means they must have mitochondria.

27 The answer is **false.**

All cells have DNA. Prokaryotes just keep theirs in their cytoplasm, in an area scientists call the *nucleoid.*

28 The answer is **b. Cell wall.**

All three other structures are only found in eukaryotes. Green bacteria do photosynthesis at their plasma membranes, not in chloroplasts.

Chapter 4

Tracking the Flow of Energy and Matter

*J*ust like you need to put gas in your car so your car can move, you need to put food in your body so that it runs. And you're not alone. Every person, as well as every other living thing, needs to "fill its tank" with matter and energy in the form of food. Cells use food molecules as building material or break them down to get the energy they need for growth and maintenance. In this chapter, I walk you through the role of food and help you break down the complicated processes that cells use to make and use food for energy and matter.

Figuring Out the Role of Food

Food molecules — in the form of proteins, carbohydrates, and fats — provide the matter and energy that every living thing needs (for more on matter and molecules, see Chapter 2). Cells use food for either matter or energy:

✔ **Organisms need matter to build their cells so they can grow, repair themselves, and reproduce.**

✔ **Organisms need energy so they can move, build new materials, and transport materials around their cells.** These activities are all examples of *cellular work,* the energy-requiring processes that occur in cells.

REMEMBER

Food is a handy package that contains two things every organism needs: matter and energy.

1. A bacterial cell absorbs sunlight and uses it to move ions across a membrane. Is the bacterial cell using the sunlight for matter or energy?

    a. Matter

    b. Energy

2. A plant cell attaches a bunch of glucose molecules to a large cellulose molecule that will become part of the plant cell wall. Is the plant cell using the glucose primarily for matter or energy?

    a. Matter

    b. Energy

Make it or break it

All organisms need food, but there's one major difference in how they approach this problem:

- ✔ **Autotrophs can make their own food.** *Auto* means "self," and *troph* means "feed," so *autotrophs* are self-feeders. Plants, algae, and green bacteria are examples of autotrophs.

- ✔ **Heterotrophs have to eat other organisms to get their food.** *Hetero* means "other," so *heterotrophs* are quite literally other-feeders. Animals, fungi, and most bacteria are examples of heterotrophs.

Although you may think that obtaining food is as easy as heading to the supermarket, pulling up to a drive-through window, or meeting the delivery guy at the front door, acquiring nutrients is actually a metabolic process. More specifically, food is made through one process and broken down through another. These processes are as follows:

- ✔ **Photosynthesis:** Only autotrophs engage in photosynthesis, a process that consists of using energy from the sun, carbon dioxide from the air, and water from the soil to make sugars. (The carbon dioxide provides the matter plants need for food-building.) When plants remove hydrogen atoms from water to use in the sugars, they release oxygen as waste.

- ✔ **Cellular respiration:** Both autotrophs and heterotrophs break food down by cellular respiration. During this process, oxygen is used to help break down food molecules such as sugars. The energy stored in the bonds of the food molecules is transferred to adenosine triphosphate (ATP; see the section "Transferring energy with ATP" later in the chapter for more info). As the energy is transferred to the cells, the matter from the food molecules is released as carbon dioxide and water.

If you think about it, photosynthesis and cellular respiration are really the opposites of each other. Photosynthesis consumes carbon dioxide and water and produces food and oxygen. Cellular respiration consumes food and oxygen and produces carbon dioxide and water. Scientists write the big-picture view of both processes as the following equations:

Photosynthesis:

$6\,CO_2 + 6\,H_2O + \text{light energy} \rightarrow C_6H_{12}O_6 + 6\,O_2$

Cellular respiration:

$C_6H_{12}O_6 + 6\,O_2 \rightarrow 6\,CO_2 + 6\,H_2O + \text{usable energy}$

Don't fall for the idea that only animals (heterotrophs) engage in cellular respiration. Autotrophs such as plants engage in it too. Think of it like this: Photosynthesis is a food-making pathway that autotrophs use to store matter and energy for later. So a plant doing photosynthesis is like you packing a lunch for yourself. You wouldn't have much point in packing the lunch if you weren't going to eat it later, right? The same is true for a plant. It does photosynthesis to store matter and energy. When it needs that matter and energy, it uses cellular respiration to "unpack" its food.

Although the equations for photosynthesis and cellular respiration are the opposite of each other from a big picture view, cellular respiration is not photosynthesis backward, and photosynthesis is not cellular respiration backward. Both photosynthesis and cellular respiration actually occur in many, many smaller steps. The steps for photosynthesis are different from the steps for cellular respiration.

3. A blue-green bacterial cell absorbs sunlight, water, and carbon dioxide and uses them to produce starch. Is the bacterium an autotroph or a heterotroph?

    a. Autotroph

    b. Heterotroph

4. Which of the following would be the most likely source of energy for a heterotroph?

    a. Sunlight

    b. Water

    c. Oxygen

    d. Food

5. Which cellular process is represented by the following reaction: $C_6H_{12}O_6 + 6 O_2 \rightarrow 6 CO_2 + 6 H_2O$ + usable energy?

    a. Cellular respiration

    b. Photosynthesis

6. What is the purpose of photosynthesis for an autotroph?

    a. To make oxygen

    b. To store matter and energy

    c. To transfer energy from food to its cells

    d. To make food for heterotrophs to eat

Feeling energized about energy

You can probably think of many kinds of energy in your life — electricity, heat, light, and chemical energy (think gasoline). Although they may seem very different, these kinds of energy represent the two main types of energy:

- **Potential energy:** This is the energy that's stored in something because of the way it's arranged or structured. Food and gasoline contain potential energy called *chemical potential energy* (energy that's stored in the bonds of molecules).

- **Kinetic energy:** This is the energy of motion. Light, heat, and moving objects all contain kinetic energy.

Energy behaves according to very specific rules that scientists call the *laws of thermodynamics*.

The *first law of thermodynamics* states that **energy can't be created or destroyed.** In other words, energy isn't created from nothing; it always comes from some other kind of energy. And when people use, say, electricity, that energy doesn't disappear. Instead, it becomes other kinds of energy, such as light or heat.

Scientists use two terms to describe energy changes:

✔ **When energy moves from one place to another, scientists say it's *transferred*.**

✔ **When energy changes from one form to another, scientists say it's *transformed*.**

7. A student says, "Plants turn energy from the sun into food." Something is wrong with this statement. What is wrong? How would you rewrite this statement to make it correct?

8. A log is burning in a campfire. The log is made mostly of the carbohydrates cellulose and lignin. As the campfire burns, campers watch the flames and feel the fire's warmth. What is happening in terms of energy as the log burns?

   a. The molecules in the wood are converted into kinetic energy (light and heat).

   b. The chemical potential energy in the wood is transformed into kinetic energy (light and heat).

   c. The kinetic energy in the wood is transformed into chemical potential energy (light and heat).

   d. The chemical potential energy in the wood is transformed into potential energy (light and heat).

9. Consider the same burning campfire log in Question 8. As the log is burning, energy transfers are taking place. Describe this transfer, stating the energy source (where the energy comes from in the example) and the energy receivers (where the energy ends up).

Getting a reaction

Living organisms follow the rules of physics and chemistry, and the human body is no exception. The first law of thermodynamics (explained in the preceding section) applies to your *metabolism,* which is all the chemical reactions that occur in your cells at one time.

Two types of chemical reactions can occur as an organism metabolizes molecules:

✔ **Anabolic reactions:** This type of reaction builds molecules. Specifically, small molecules are combined into large molecules for repair, growth, or storage.

✔ **Catabolic reactions:** This type of reaction breaks down molecules to release their stored energy.

During chemical reactions, atoms receive new bonding partners, and energy may be transferred. As an example, take another look at the summary reaction for cellular respiration:

$$C_6H_{12}O_6 + 6\,O_2 \rightarrow 6\,CO_2 + 6\,H_2O + \text{usable energy}$$

✔ The arrow in the middle of this chemical equation represents the reaction; in other words, the actual changes that occur in the molecules.

✔ The molecules to the left of the arrow are the *reactants,* the molecules that enter into the reaction — in this case, glucose ($C_6H_{12}O_6$) and oxygen (O_2).

✔ The molecules to the right of the arrow are the *products*, the molecules that leave the reaction — in this case, carbon dioxide (CO_2) and water (H_2O).

✔ Glucose and oxygen have more chemical potential energy than do carbon dioxide and water, so when this reaction happens, some of the original stored energy becomes available to the cell. This reaction is an example of a catabolic reaction because it breaks down a larger molecule into smaller molecules and makes energy available to cells.

✔ Notice that the number of each type of atom is the same on both sides of the arrow. That's because in living things, matter is never created or destroyed — a rule that's sometimes called the *law of conservation of matter*. If you count the number of carbon atoms, for example, you find that the glucose molecule has 6 (C_6) and the 6 carbon dioxide molecules have 6 (6×1 for each CO_2). Likewise, the glucose has 12 hydrogen atoms and the 6 molecules of water have 12 hydrogen atoms (6×2 for each molecule). For oxygen, the reactants have a total of 18 atoms (6 in glucose and 6×2 for the 6 oxygen molecules), as do the products (6×2 for the 6 CO_2 molecules and 6 in the water molecules). The same number of atoms that entered into this chemical reaction with the reactants left again with the products — they just had new molecular partners.

When a chemical equation is written so that the number and types of atoms in the reactants are equal to the number and types of atoms in the products, chemists say the equation is *balanced*.

Q. The following reaction shows the combustion (burning) of propane in the presence of oxygen to produce water and carbon dioxide. The number of carbon atoms in a molecule of propane has been left out. How many carbon atoms should one molecule of propane have in order to balance this equation?

$C\_H_8 + 5\ O_2 \rightarrow 3\ CO_2 + 4\ H_2O$ + usable energy

A. The number of carbon atoms must be the same on both sides of the reaction. So, if you look at the products, you see that the only carbon atoms are those in carbon dioxide. Three molecules of carbon dioxide are produced for each molecule of propane, so the answer is 3.

$C_3H_8 + 5\ O_2 \rightarrow 3\ CO_2 + 4\ H_2O$ + usable energy

Now you try to balance some equations:

10. The following reaction shows the formation of methane (CH_4) from carbon dioxide (CO_2) and hydrogen gas (H_2). How many molecules of hydrogen gas are required to balance this equation?

$CO_2 + \_\ H_2 \rightarrow CH_4 + 2\ H_2O$

a. 1

b. 2

c. 3

d. 4

11. The following reaction shows the oxidation of ammonia (NH_3) into nitric oxide (NO) and water (H_2O). How many molecules of oxygen gas (O_2) are required to balance this equation?

$$4\ NH_3 + \_\ O_2 \rightarrow 4\ NO + 6\ H_2O$$

a. 1

b. 3

c. 4

d. 5

Transferring energy with ATP

Cells transfer energy between anabolic and catabolic reactions by using an energy carrier called *adenosine triphosphate* (or ATP for short). Energy from catabolic reactions is transferred to ATP, which then provides energy for anabolic reactions. ATP has three phosphates attached to it (tri- means "three," so triphosphate means "three phosphates"). When ATP supplies energy to a process, one of its phosphates gets transferred to another molecule, turning ATP into adenosine diphosphate (ADP). Cells re-create ATP by using energy from catabolic reactions to reattach an inorganic phosphate group (P_i) to ADP. Cells constantly build and break down ATP, creating the ATP/ADP cycle shown in Figure 4-1.

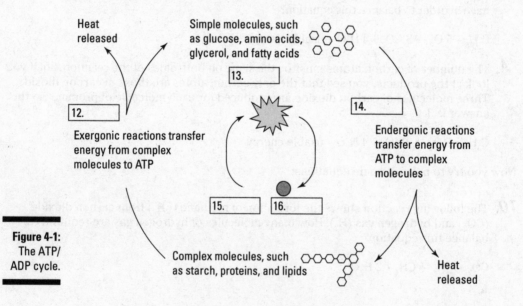

Heat released

Simple molecules, such as glucose, amino acids, glycerol, and fatty acids

13.

12.

Exergonic reactions transfer energy from complex molecules to ATP

14.

Endergonic reactions transfer energy from ATP to complex molecules

15.   16.

Complex molecules, such as starch, proteins, and lipids

Heat released

Figure 4-1:
The ATP/
ADP cycle.

Cells have large molecules, such as fats and complex carbohydrates, that contain stored energy, but when cells are busy doing work, they need a handy source of energy. That's where ATP comes in. Cells keep ATP on hand to supply energy for all the work of the cell.

Think of ATP like cash in your pocket. You may have money in the bank, but that money isn't always easy to get, which is why you keep some cash in your pocket to quickly buy what you need. After you spend all your cash, you have to go back to the bank or an ATM to get more. For living things, the energy stored in large molecules is like money in the bank. Cells break down ATP just like you spend your cash. Then, when cells need more ATP, they have to go back to the bank of large molecules and break down some more.

12.–16. Use the terms that follow to label the ATP/ADP cycle in Figure 4-1.

  a. ATP

  b. ADP

  c. P_i

  d. Catabolism

  e. Anabolism

17. Which molecules have more stored energy?

  a. ADP + P_i

  b. ATP

Moving your metabolism with enzymes

The pace of life is so fast that cells can't just wait around for chemical reactions to happen; they have to make them happen quickly. Fortunately, cells have the perfect tool at their disposal in the form of proteins called *enzymes* (see Figure 4-2). Enzymes make it easier for reactants to interact with each other, speeding up the rate of reactions so that cells can get the matter and energy they need to grow and move. Because enzymes speed up reactions, scientists say that enzymes act as biological *catalysts.* Enzymes are incredibly specific, so every chemical reaction in cells requires its own enzyme.

Enzymes are specific because they're folded into unique shapes that have pockets, called *active sites,* that they use to attach to reactants. Scientists have a special name, *substrates,* for the reactants in enzyme-catalyzed reactions (see Figure 4-2). Only certain substrates fit properly into the active site of a particular enzyme.

Right now, in every one of your cells, literally hundreds of chemical reactions are taking place, and each reaction requires a special enzyme. Without the right enzyme, a reaction can't happen. And if one reaction can't happen, then other reactions connected to it will also be blocked. The result of an uncompleted metabolic process is the lack of a particular metabolic product. Without a needed product, a function can't be performed, which negatively affects the organism and may result in disease.

Q. A cell has a metabolic pathway that contains the following reactions:

$$O \rightarrow P \rightarrow Q \rightarrow R \rightarrow S$$

How many different enzymes would the cell need to produce product S?

A. The pathway has four reaction arrows, so it consists of four reactions. Each reaction requires a specific enzyme, so the cell would need four enzymes.

The shape of an enzyme is so important to its job that if something causes the enzyme to unfold, it won't work anymore, and the reaction it catalyzes will stop. Three environmental conditions can cause enzymes to unfold, or *denature*.

- ✔ **High temperatures:** Every enzyme has a temperature range at which it functions best. For example, human enzymes typically work best at a human body temperature of 98 degrees F (37 degrees C). If temperatures around a cell become higher than its optimum range, enzymes may denature and the cell may die.

- ✔ **Changes in pH:** Every enzyme also has a range of pH values at which it works best. The cytoplasm of human cells is around pH 7, so most human cellular enzymes work best at this pH.

- ✔ **Exposure to salts:** As salts dissolve, they release ions. The positive and negative charges of the ions can interfere with the weak bonds that hold enzymes together.

Question 18 refers to the following metabolic pathway in a cell:

$$J \rightarrow K \rightarrow L \rightarrow M \rightarrow N \rightarrow O \rightarrow P$$

18. How many enzymes would the cell require to complete this pathway?

   a. One

   b. Three

   c. Six

   d. Seven

Questions 19 and 20 refer to Figure 4-2.

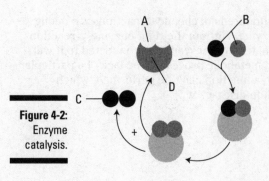

Figure 4-2:
Enzyme
catalysis.

19. Which letter in Figure 4-2 represents the product of the reaction?

　a. A

　b. B

　c. C

　d. D

20. Which letter in Figure 4-2 represents the active site of the enzyme?

　a. A

　b. B

　c. C

　d. D

21. Explain why a high fever can cause a person to die. In your explanation, be sure you give details about what's happening to the molecules in the person's cells.

Reduce, reuse, recycle

Reactions change substrates into products, but they recycle the enzymes; that is, enzymes are the same at the end of a reaction as they were at the beginning, and they can do their job again. For example, the first enzymatic reaction discovered was the one that breaks down the waste product urea into products that the body can excrete. The enzyme *urease* catalyzes the reaction between the reactants urea and water, yielding the products carbon dioxide and ammonia, which the body can easily excrete.

urease

Urea + water ↔ carbon dioxide + ammonia

In this reaction, the enzyme urease helps the substrates, urea and water, combine with each other. The bonds between the atoms in urea and water break and then re-form between different combinations of atoms, forming the products carbon dioxide and ammonia. When the reaction is over, urease is unchanged and can catalyze another reaction between urea and water.

The names of most enzymes end with the suffix –ase.

On their own, reactants could occasionally collide with each other the right way to start a reaction. But they wouldn't do it nearly often enough to keep up with the fast pace of life in a cell. Without enzymes, your body wouldn't be able to, say, get rid of urea fast enough, leading to a toxic buildup of urea. That's where the enzyme urease comes into play. It binds the substrates in its active site and brings them together in a way that requires less energy for them to react.

Because enzymes decrease the amount of energy that substrates need to react with each other, scientists say that enzymes lower the *activation energy* needed to start a reaction.

Whatever you do, don't fall for the idea that enzymes add energy to reactions to make them happen. They don't. In fact, they don't add *anything* to a reaction; they just help the reactants get together in the right way, lowering the barrier to the reaction. In other words, enzymes don't add energy; they just make it so the reactants have enough energy on their own.

Cells regulate most of their metabolic pathways by controlling enzymes with *feedback inhibition,* which relies on the product of the pathway. If a pathway produces a product faster than the cell can use it, the product builds up in the cell. During feedback inhibition, the product binds to a special regulatory site, called an *allosteric site,* on one of the initial enzymes in the pathway, changing the enzyme's shape so that it no longer works properly. After an initial enzyme is shut down, the entire pathway stops, and the cell stops producing the product. One of the cool things about feedback inhibition is that it's reversible; as the cell uses up the product it has on hand, the product no longer binds to the enzyme, and the pathway starts up again.

22. Which of the following statements best explains how enzymes speed up chemical reactions?

   a. Enzymes make more substrate so there's more available for reactions to occur.

   b. Enzymes add activation energy to the reaction, making it possible for reactions to occur.

   c. Enzymes create a situation that enables substrates to react with their available energy.

   d. Enzymes turn into products, making more products available.

23. Given the following metabolic reaction in a cell:

   $X \rightarrow Y \rightarrow Z \rightarrow A$

   Which molecule in this pathway would most likely act as an inhibitor of the entire pathway?

   a. Molecule X

   b. Molecule Y

   c. Molecule Z

   d. Molecule A

Photosynthesis: Cooking Up Carbohydrates

Autotrophs combine matter and energy to make food in the form of carbohydrates. With those carbohydrates, plus some nitrogen and minerals from the soil, autotrophs can make all the types of molecules they need to build their cells. The chemical formula for *glucose,* the most common type of simple sugar found in cells, is $C_6H_{12}O_6$. To build glucose, autotrophs need carbon, hydrogen, and oxygen atoms, plus energy to combine them into sugar.

   ✔ The carbon and oxygen for the sugars come from carbon dioxide in the Earth's atmosphere.

   ✔ The hydrogen for the sugars comes from water in the soil.

   ✔ The energy to build the sugars comes from the sun (but only in autotrophs that use photosynthesis).

A common misconception that many people have is that plants get the matter they need to grow from the soil. This seems like a perfectly logical idea, given that plants grow with their roots stuck in the ground. However, Belgian scientist Jean Baptiste van Helmont showed that a tree that gained approximately 169 pounds in mass took only 2 ounces of dry material from the soil (not counting water). This experiment proved that plants don't take lots of material from the soil. Instead, they get most of the matter they need to grow from the carbon dioxide in the air. Plants collect a lot of carbon dioxide molecules (CO_2) and combine them with water molecules (H_2O) to build sugars such as glucose ($C_6H_{12}O_6$). From the soil, plants take only water and some small amounts of minerals like nitrogen.

If you've ever seen a piece of dry ice, you've seen carbon dioxide in its solid form. So even though carbon dioxide gas in the air seems like nothing, if you pack the atoms more tightly together, they become solid. Imagine autotrophs as cells that gather carbon dioxide gas (CO_2) and pack the atoms tightly, along with those from water (to provide the H), forming the solid material that organisms use to build their cells.

Photosynthesis occurs in two main steps (Figure 4-3 depicts both in action):

1. **The light reactions of photosynthesis transform light energy into chemical energy.**

   The chemical energy is stored in the energy carrier ATP. During the energy transformation, cells remove electrons from water (H_2O) molecules, ultimately leading to the separation of hydrogen atoms (H) from water molecules and the production of oxygen gas (O_2) as waste.

2. **The light-independent reactions of photosynthesis produce food.**

   ATP from the light reactions supplies the energy needed to combine carbon dioxide (CO_2) and hydrogen atoms from water to make glucose ($C_6H_{12}O_6$).

24. Plant cell walls contain lots of *cellulose,* a polysaccharide made up of carbon, hydrogen, and oxygen atoms. Where do plants get the carbon atoms they need to build cellulose?

   a. From the air

   b. From the soil

   c. From water

25. Where do plants get the hydrogen atoms they need to build cellulose?

   a. From the air

   b. From the soil

   c. From water

26. Like all cells, plant cells must make lots of proteins. Proteins contain the atoms carbon, hydrogen, oxygen, nitrogen, and sulfur. Where do plants get the nitrogen atoms they need to build proteins?

   a. From the air

   b. From the soil

   c. From water

Light reactions of photosynthesis: Transforming energy from the ultimate energy source

The sun is a perfect energy source, a nuclear reactor positioned at a safe distance from Earth. It contains all the energy you could ever need . . . if only you could capture it. Well, green bacteria figured out how to do just that more than 2.5 billion years ago, which shows that photosynthetic autotrophs are way ahead of humans on this one.

Plants, algae, and green bacteria use *pigments* to absorb light energy from the sun. You're probably most familiar with the pigment *chlorophyll,* which colors the leaves of plants green. The chloroplasts in plant cells contain lots of chlorophyll in their membranes so they can absorb light energy (see Chapter 3 for more on chloroplasts).

During the light reactions of photosynthesis, chloroplasts absorb light energy from the sun and transform it into the chemical energy stored in ATP. During this energy transformation, the chloroplasts make two products that are needed by plants for making sugars:

- ✔ **ATP:** Energy from the sun is stored in ATP so that the energy can be used to build sugars.

- ✔ **Nicotinamide adeninine dinucleotide phosphate (NADPH):** Just like ATP is an energy carrier, NADPH is an *electron carrier.* When chloroplasts absorb light energy, they split water molecules. The electrons from the water molecules transfer to the "empty" form of the carrier, $NADP^+$, turning it into the "full" form of the carrier, NADPH. The electrons move as part of hydrogen atoms, so you can also call NADPH a *hydrogen carrier.* NADPH takes the electrons, as hydrogen atoms, to the light-independent reactions, where the cell uses them to build sugars. As a byproduct of the splitting of water, plants produce the oxygen (O_2) that you breathe.

Test your knowledge of the light reactions by answering these questions:

27. Which of the following accurately describes the light reactions of photosynthesis?

    a. Cells transform kinetic energy (light) into chemical potential energy.

    b. Cells build glucose molecules out of carbon dioxide and water.

    c. Cells transform potential energy (light) into chemical potential energy.

    d. Cells build glucose molecules out of light.

28. The products of the light reactions of photosynthesis are ATP, oxygen gas (O_2), and electrons carried in an electron carrier called NADPH. Where do the oxygen molecules produced by plants come from?

    a. From the oxygen in the air (O_2)

    b. From carbon dioxide (CO_2)

    c. From water (H_2O)

29. Describe the energy transfer that occurs during the light reactions of photosynthesis. In your answer, be sure to name the original source of the energy and the final receiver of the energy (where the energy ends up at the end of the light reactions).

30. A classmate says, "Plants make oxygen so we can breathe." This statement implies that plants make oxygen because they want to help humans. How would you rephrase this statement to make it more accurate from the point of view of plants?

Light-independent reactions of photosynthesis: Putting matter and energy together

In the light-independent reactions, plants combine carbon dioxide molecules and hydrogen atoms from water molecules, creating glucose molecules. To make glucose, plants first take carbon dioxide out of the air through a process called *carbon fixation* (taking carbon dioxide and attaching it to a molecule inside the cell). They then use the energy from the ATP and the electrons carried as part of hydrogen atoms by NADPH (that originally came from water) to convert the carbon dioxide to sugar.

Some of the products of the light-independent reactions return to participate again in the light reactions:

✔ As plants transfer energy from ATP to carbohydrates, the ATP breaks down to ADP and P_i.

✔ As plants transfer electrons from NADPH to carbohydrates, the NADPH converts back to $NADP^+$.

The light-independent reactions form a metabolic cycle that's also known as the *Calvin-Benson cycle* (named after the scientists who discovered it).

 As their name indicates, the light-independent reactions of photosynthesis don't need direct sunlight to occur. However, plants need the products of the light reactions to run the light-independent reactions, so really, the light-independent reactions can't happen if the light reactions don't happen.

When plants have made more glucose than they need, they store their excess matter and energy by combining glucose molecules into larger carbohydrate molecules, such as starch. When necessary, plants can break down the starch molecules to retrieve glucose for energy or to create other compounds, such as proteins and nucleic acids (with added nitrogen taken from the soil) or fats (many plants — such as olives, corn, peanuts, and avocados — store matter and energy in oils).

31. Which of the following accurately describes the light-independent reactions of photosynthesis?

a. Cells transform kinetic energy (light) into chemical potential energy.

b. Cells transfer chemical potential energy (in ATP) into chemical potential energy (in carbohydrates).

c. Cells transform kinetic energy (in ATP) into chemical potential energy (in carbohydrates).

d. Cells build glucose molecules out of light.

32. True or false: During the light-independent reactions of photosynthesis, plant cells take in carbon dioxide and produce oxygen.

33. Which of the following has more electrons (as part of hydrogen atoms)?

  a. NADP⁺

  b. NADPH

34.–45. Use the terms that follow to label the main events of photosynthesis in Figure 4-3.

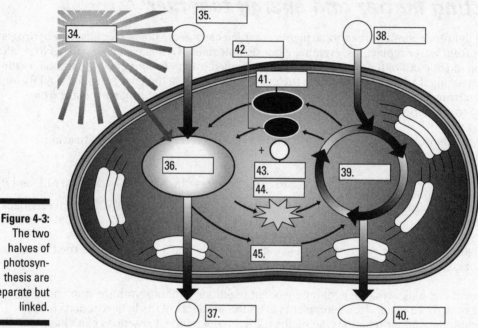

Figure 4-3:
The two halves of photosynthesis are separate but linked.

  a. CO_2

  b. H_2O

  c. O_2

  d. Carbohydrates

  e. Light reactions

  f. Light-independent reactions

  g. ADP

  h. P_i

  i. ATP

  j. Light

  k. NADP⁺

  l. NADPH

46. The light-independent reactions of photosynthesis can be thought of as a matter transfer because matter from the environment is transferred into plant cells. Describe this matter transfer, naming the environmental source of the matter and the receiver of the matter (where the matter ends up at the end of the light-independent reactions).

Questions 47 and 48 refer to the following scenario:

Indian Pipe is a completely white plant that doesn't make chlorophyll. It can grow in very shaded forest areas that don't have much light for photosynthesis. This plant grows very near other plants and is actually connected to them by fungi (like mushrooms) that grow in the soil. When scientists traced the carbon atoms that make up the molecules in Indian Pipe, they found that the carbon atoms actually come from the nearby trees in sugars that pass from the tree, through the fungi, and into the Indian Pipe.

47. Does Indian Pipe do the light reactions of photosynthesis?

    a. Yes, it's a plant so it does all of photosynthesis.

    b. No, it can't because it doesn't have chlorophyll.

48. Is Indian Pipe an autotroph or a heterotroph?

    a. Autotroph

    b. Heterotroph

Cellular Respiration: Extracting Energy from Food

Autotrophs and heterotrophs break down food to transfer energy from food to ATP. Many cells break down food using a collection of metabolic pathways called *cellular respiration*. The cells of animals, plants, and many bacteria use oxygen to help with the energy transfer during cellular respiration; in these cells, cellular respiration is also referred to as *aerobic respiration* (*aerobic* means "with air").

Cellular respiration is different from respiration, which is more commonly referred to as breathing. Cellular respiration is what happens inside cells when they use oxygen to transfer energy from food to ATP.

Three separate pathways combine to form the process of cellular respiration. The first two, *glycolysis* and the *Krebs cycle,* break down food molecules. The third pathway, *oxidative phosphorylation,* transfers the energy from the food molecules to ATP. Here are the basics of how cellular respiration works:

✔ During glycolysis, which occurs in a cell's cytoplasm, cells break down glucose into *pyruvate,* a three-carbon compound. After glycolysis, pyruvate is broken down into a two-carbon molecule called *acetyl-coA.*

✔ After pyruvate is converted to acetyl-coA, cells use the Krebs cycle (which occurs inside the mitochondrion) to break down acetyl-coA into carbon dioxide.

✔ During *oxidative phosphorylation,* which occurs in the mitochondrion's inner membrane, cells transfer energy from the breakdown of food to ATP.

The Krebs cycle is also called the *citric acid cycle* and the *tricarboxylic acid (TCA) cycle.*

Do you know the three pathways of cellular respiration? Try these questions to check:

49. Which pathway of cellular respiration produces pyruvate?

a. Glycolysis

b. Krebs cycle

c. Oxidative phosphorylation

d. Light-independent reactions

50. Which pathway of cellular respiration occurs in the cytoplasm?

a. Glycolysis

b. Krebs cycle

c. Oxidative phosphorylation

d. Light-independent reactions

51. Which pathway of cellular respiration produces the most ATP?

a. Glycolysis

b. Krebs cycle

c. Oxidative phosphorylation

d. Light-independent reactions

Glycolysis and the Krebs cycle: Breaking down glucose to carbon dioxide

In order to transfer energy from food to ATP, all organisms must first break down the large molecules in food into their smaller subunits. In you, this happens during digestion. After cells break food molecules down into their subunits, the small molecules can be further broken down to transfer their energy to ATP. During cellular respiration, enzymes slowly rearrange the atoms in the small molecules. Each rearrangement produces a new molecule that may be useful to the cell. In addition, some reactions allow the transfer of energy, electrons, or carbon atoms:

✔ **Cells release energy that can be transferred to ATP.** Cells quickly use this ATP for cellular work, such as movement and building new molecules.

✔ **Cells oxidize food molecules and transfer electrons and energy to electron carriers.** *Oxidation* is the process that removes electrons from molecules; *reduction* is the process that gives electrons to molecules. During cellular respiration, enzymes remove electrons from food molecules and transfer the electrons to the electron carriers *nicotinamide adenine dinucleotide* (NAD⁺) and *flavin adenine dinucleotide* (FAD). NAD⁺ and FAD receive the electrons as part of hydrogen (H) atoms, which changes them to their reduced forms NADH and FADH₂. Next, NADH and FADH₂ donate the electrons to the process of oxidative phosphorylation, which transfers energy to ATP.

✔ **Cells release carbon dioxide (CO_2).** Cells return CO_2 to the environment as waste, which is great for the autotrophs that require CO_2 to produce the food heterotrophs will eventually eat. For every glucose molecule ($C_6H_{12}O_6$) that cells break down completely by cellular respiration, all six carbon atoms from glucose leave the cell as waste molecules of CO_2.

Just like $NADP^+$ in photosynthesis, NAD^+ and FAD act like electron shuttle buses for the cell. The empty buses, NAD^+ and FAD, drive up to oxidation reactions and collect electron passengers. When the electrons get on the bus, the driver puts up the sign, *H*, to show that the bus is full. Then, the full buses, NADH and $FADH_2$, drive over to reactions that need electrons and let the passengers off. The buses are empty again, and they drive to another oxidation reaction to collect more passengers. During cellular respiration, the electron shuttle buses drive a loop between the reactions of glycolysis and the Krebs cycle, where they pick up passengers, to the electron transport chain, where they drop off their passengers.

Following is a summary of how different molecules break down in the first two pathways of cellular respiration:

✔ During glycolysis, glucose breaks down into two molecules of pyruvate. The backbone of glucose has six carbon atoms, whereas the backbone of pyruvate has three carbon atoms. During glycolysis, energy transfers result in a net gain of two ATP and two molecules of the reduced form of the coenzyme NADH.

✔ Pyruvate converts to acetyl-coA, which has two carbon atoms in its backbone. One carbon atom from pyruvate is released from the cell as CO_2, and electrons from pyruvate transfer to NAD^+. The conversion of pyruvate to acetyl-coA happens twice for every glucose molecule that enters cellular respiration.

✔ During the Krebs cycle, each acetyl-coA breaks down into two carbon dioxide molecules (CO_2). Energy transfers during the Krebs cycle produce an additional three molecules of NADH, one molecule of $FADH_2$, and one molecule of ATP. The Krebs cycle happens twice for every glucose molecule that enters cellular respiration.

52. True or false: Glycolysis releases CO_2 molecules to the atmosphere.

   a. True

   b. False

53. Which form of the electron carrier is more reduced (is carrying electrons)?

   a. NAD^+

   b. NADH

54. True or false: Electrons transfer to electron carriers during both glycolysis and the Krebs cycle.

   a. True

   b. False

55. True or false: Cells transfer energy to ATP during both glycolysis and the Krebs cycle.

   a. True

   b. False

Oxidative Phosphorylation: Transferring energy to ATP

In the mitochondrial cristae (see Chapter 3 for details on mitochondrial structure), hundreds of little cellular machines are busily working to transfer energy from food molecules to ATP. The cellular machines are called *electron transport chains,* and they're made of a team of proteins that sit in the membranes and transfer energy and electrons throughout the machines.

The electron carriers NADH and $FADH_2$ are the source of energy and electrons to the electron transport chain. These carriers stored the energy and electrons during the catabolic reactions of glycolysis and the Krebs cycle. NADH and $FADH_2$ transfer energy and electrons to the proteins of the electron transport chain, then return to the reactions of glycolysis and Krebs to collect more. As the proteins of the electron transport chain transfer the electrons through the chain, the energy and electrons each transfer along a unique path:

✔ **Electrons transfer from NADH (or $FADH_2$) to the proteins of the electron transport chain to oxygen.** Oxygen collects the electrons at the end of the chain. (If you didn't have oxygen around at the end of the chain to collect the electrons, no energy transfer could occur.) When oxygen accepts the electrons, it also picks up protons (H^+) and becomes water (H_2O).

✔ **Energy transfers from NADH (or $FADH_2$) to the proteins of the electron transport chain to a proton gradient to ATP.** The electron transport proteins use energy from the transfer of electrons to move protons (H^+) across the inner membrane of the mitochondrion. They pile up the protons like water behind the "dam" of the inner membrane, storing potential energy. These protons then flow back across the inner membrane through a protein, called *ATP synthase,* that transforms the kinetic energy from the moving protons to chemical energy in ATP by capturing the energy in chemical bonds as it adds phosphate molecules to ADP.

The entire process of how ATP is made at the electron transport chain is called the *chemiosmotic theory of oxidative phosphorylation* (check out the diagram in Figure 4-4).

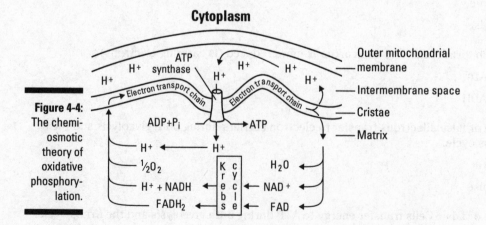

Figure 4-4: The chemi-osmotic theory of oxidative phosphory-lation.

At the end of the entire process of cellular respiration, the energy transferred from glucose is stored in 36 to 38 molecules of ATP, which are available to be used for cellular work.

56. Take out two colored pencils or highlighters. On Figure 4-4, use one color to mark the pathway for energy during oxidative phosphorylation. Draw a box around the source of energy to the electron transport chain and a circle around the final receiver of the energy. Use the second color to trace the pathway traveled by electrons during oxidative phosphorylation. Again, draw a box around the source of electrons and a circle around the final receiver of the electrons.

57. During cellular respiration, cells remove electrons from glucose. Where do these electrons ultimately end up at the end of the process?

   a. ATP

   b. NADH

   c. H_2O

   d. CO_2

58. During cellular respiration, cells transfer energy from glucose. Where does this energy ultimately end up at the end of the process?

   a. ATP

   b. NADH

   c. H_2O

   d. CO_2

59. Which of the following pathways correctly shows the movement of electrons from glucose during the process of cellular respiration?

   a. Glucose → NADH → ATP

   b. Glucose → ATP → CO_2

   c. Glucose → NADH → $FADH_2$

   d. Glucose → NADH → H_2O

60.–62. Use the terms that follow to label the major pathways of cellular respiration shown in Figure 4-5.

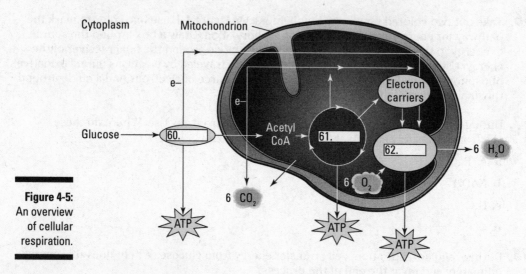

    a. Krebs cycle

    b. Glycolysis

    c. Oxidative phosphorylation

63.–67. Cells continuously transfer energy by making and breaking ATP, creating the ATP/ADP cycle shown in Figure 4-1. Similarly, cells constantly transfer electrons by oxidizing and reducing electron carriers, creating cycles like the NAD^+/NADH cycle, the FAD/$FADH_2$ cycle, and the $NADP^+$/NADPH cycle. Use the terms that follow to identify which form of each carrier would perform the task described in the following scenarios. For some scenarios, more than one answer is possible.

    a. NAD^+

    b. NADH

    c. FAD

    d. $FADH_2$

    e. $NADP^+$

    f. NADPH

63. Accepts electrons during the Krebs cycle.

64. Accepts electrons during the light reactions of photosynthesis.

65. Donates electrons during oxidative phosphorylation.

66. Donates electrons during the light-independent reactions of photosynthesis.

67. Accepts electrons during glycolysis.

Answers to Questions on Tracking the Flow of Energy and Matter

The following are answers to the practice questions presented in this chapter.

 1 The answer is **b. Energy.**

Sunlight isn't made of atoms, so it has no mass (weight) and therefore isn't matter.

 2 The answer is **a. Matter.**

Although glucose does contain stored energy, in this example the plant is using the glucose as building material, so the primary use is matter.

3 The answer is **a. Autotroph.**

This cell is making food.

4 The answer is **d. Food.**

Heterotrophs can't make their own food so they don't usually use sunlight for energy. Water and oxygen don't contain very much stored energy (for example, water has zero calories).

 5 The answer is **a. Cellular respiration.**

6 The answer is **b. To store matter and energy.**

Although photosynthesis does produce oxygen, to the autotroph, oxygen is a waste product. For autotrophs, the purpose of photosynthesis is to transfer energy and matter from the environment into food for their own use later on.

7 The answer is **partly correct** because plants do store energy from the sun in food molecules.

However, the student is making a pretty big mistake. Cells can't turn energy into matter, so the student is forgetting that plants need actual atoms to build food molecules. Those atoms come from carbon dioxide and water molecules. It would be much more accurate to say that the energy from the sun lets plant cells do the work necessary to put the carbon dioxide and water together to build food.

8 The answer is **b. The chemical potential energy in the wood is transformed into kinetic energy (light and heat).**

Cells can't convert molecules into energy; they can only transfer energy from one type to another. Chemical energy in molecules is potential energy. Light and heat are both forms of kinetic energy.

9 The log is the source of energy because chemical potential energy is stored in the molecules that make up the wood. The energy is transferred to the environment (receiver) as heat and light.

10 The answer is **d. 4.**

The products have 4 atoms of hydrogen in methane (CH_4) and another 4 in the 2 molecules of water (H_2O). So the reaction requires 4 molecules of hydrogen gas (H_2).

11 The answer is **d. 5.**

12–**16** The following is how Figure 4-1 should be labeled:

12 **d. Catabolism**; 13 **a. ATP**; 14 **e. Anabolism**; 15 **b. ADP**; 16 **c. P**$_i$

17 The answer is **b. ATP.**

18 The answer is **c. Six.**

Six arrows means six chemical reactions, and in a cell, that means six enzymes.

19 The answer is **c. C.**

20 The answer is **a. A.**

21 High body temperatures raise the temperature of cells above the optimal range for cellular enzymes. The enzymes denature, or unfold, preventing them from working properly. If enzymes don't work properly, then cells don't work properly. Cells can't get the energy they need to survive or make the molecules they need to function, so they die. When heart cells die, the heart stops functioning. When brain cells die, the brain stops functioning. So basically, when cells die, organisms die.

22 The answer is **c. Enzymes create a situation that enables substrates to react with their available energy.**

Enzymes lower the activation energy for the reaction so substrates can more easily react with each other with the amount of energy they already have. Enzymes don't make substrates; they make products. But they themselves aren't changed during the reaction. And enzymes don't provide energy for reactions to happen.

23 The answer is **d. Molecule A.**

Cells often use the end product of a pathway as an inhibitor of the pathway in a process called *feedback inhibition.*

24 The answer is **a. From the air.**

Plants take in carbon dioxide and use it to build their molecules.

25 The answer is **c. From water.**

Carbon dioxide from the air provides carbon and oxygen atoms to plants; water from the soil provides hydrogen atoms.

26 The answer is **b. From the soil.**

Plants take minerals like nitrogen from the soil.

27 The answer is **a. Cells transform kinetic energy (light) into chemical potential energy.**

The light reactions transfer kinetic energy from the sun into the chemical energy of ATP.

28 The answer is **c. From water (H$_2$O).**

When plants take hydrogen atoms for glucose from water, they release the oxygen atoms as waste.

29 The source of energy is the sun. The light reactions of photosynthesis transfer energy from the sun to the plant cells (the receiver) by transforming the light energy to chemical potential energy in ATP.

30 Plants make oxygen as a waste product of photosynthesis. Plants split water molecules, taking their electrons for photosynthesis. When H_2O molecules are split, the oxygen atoms are released as O_2. Animals (and other cells) use the O_2, but to the plants, the point of the light reactions is to capture energy from the sun into a usable form for cells. The point of the entire process of photosynthesis is to store matter and energy by making food molecules.

31 The answer is **b. Cells transfer chemical potential energy (in ATP) into chemical potential energy (in carbohydrates).**

32 The answer is **false.**

Plants do take in CO_2 during the light-independent reactions, but O_2 is only produced during the light reactions.

33 The answer is **b. NADPH.**

NADPH is the "full" form of the bus that's carrying electrons.

34 – 45 The following is how Figure 4-3 should be labeled:

34 **j. Light**; 35 **b. H_2O**; 36 **e. Light reactions**; 37 **c. O_2**; 38 **a. CO_2**; 39 **f. Light-independent reactions**; 40 **d. Carbohydrates**; 41 **k. NADP+**; 42 **g. ADP**; 43 **h. P_i**; 44 **i. ATP**; 45 **l. NADPH**

46 The light-independent reactions of photosynthesis transfer matter from the environment (the source) to plant cells (the receiver). The matter comes from the environment in the form of CO_2 from the atmosphere and H_2O from the soil. It's converted into carbohydrates during the light-independent reactions.

47 The answer is **b. No, it can't because it doesn't have chlorophyll.**

Chlorophyll is the green pigment that plants need to absorb light energy. Without chlorophyll (Indian Pipe is totally white, not green), Indian Pipe can't absorb light energy.

48 The answer is **b. Heterotroph.**

Indian Pipe gets its carbon from its neighbors, so even though it's a plant, it's acting like a heterotroph!

49 The answer is **a. Glycolysis.**

50 The answer is **a. Glycolysis.**

51 The answer is **c. Oxidative phosphorylation.**

52 The answer is **b. False.**

Cells produce CO_2 during the oxidation of pyruvate to acetyl-coA and during the Krebs cycle.

53 The answer is **b. NADH.**

NAD+ accepts the electrons, but after it does it becomes NADH. You can remember this by looking at the H in the name. If the shuttle bus is showing the H, that means it's carrying passengers.

54 The answer is **a. True.**

Every stage of cellular respiration transfers some electrons to electron carriers, although the Krebs cycle definitely transfers the most.

55 The answer is **a. True.**

A little bit of energy transfers to ATP during glycolysis and the Krebs cycle.

56 For the path of energy, you should've drawn a box around NADH and $FADH_2$ as the source and then drawn a line from these molecules across the cristae and into the intermembrane space (where the H^+ are). From the H^+, your energy line should go through the tunnel that represents ATP synthase and then over to the ATP molecule. The ATP molecule should be circled as the final receiver of the energy.

For the path of electrons, you should again begin with a box around NADH and $FADH_2$ as the source. From these molecules, your line should go through the electron transport chain and to the H_2O. The H_2O should be circled as the final destination of the electrons. (Oxygen accepts the electrons but is converted to water in the process.)

57 The answer is **c. H_2O.**

Electrons transfer through the electron transport chain until they get picked up by oxygen (O_2). When oxygen picks up the electrons, it also grabs a couple of protons (H^+), and everything combines together to form water (H_2O).

58 The answer is **a. ATP.**

The whole point of cellular respiration is to transfer energy from food into ATP.

59 The answer is **d. Glucose $\rightarrow$ NADH $\rightarrow$ H_2O.**

60 – 62 The following is how Figure 4-5 should be labeled:

60 **b. Glycolysis;** 61 **a. Krebs cycle;** 62 **c. Oxidative phosphorylation**

63 The answer is **a. NAD^+** and **c. FAD.**

64 The answer is **e. $NADP^+$**

65 The answer is **b. NADH** and **d. $FADH_2$.**

66 The answer is **f. NADPH.**

67 The answer is **a. NAD^+.**

Part II
Creating the Future with Cell Division and Genetics

The 5th Wave   By Rich Tennant

In this part . . .

Today, scientists know more about the mysteries of DNA than ever before and have figured out ways to harness its power for medicine, agriculture, and technology. But besides being the acronym of *deoxyribonucleic acid,* what exactly *is* DNA?

DNA is a code that contains the blueprints for building the molecules that do the work of cells. Cellular function determines the traits of organisms, so ultimately, DNA determines these traits. When living things reproduce, they give copies of their DNA to their offspring, passing their traits from one generation to the next. All of your traits — everything from your eye color to your foot size — come from your family tree.

In this part, I explain how cells reproduce and how they read the code of DNA to build proteins. I also introduce you to some of the techniques that scientists use to read and work with DNA. You benefit from many chances to test your knowledge along the way.

Chapter 5

Divide and Conquer: Recognizing the Phases of Cell Division

. . . .

In This Chapter

▶ Understanding the basics of cell reproduction

▶ Copying your DNA

▶ Following the phases of mitosis

▶ Reproducing sexually through meiosis

. . . .

All living things can reproduce their cells for growth, repair, and reproduction. Asexual reproduction by *mitosis* creates cells that are identical to the parent cell. Sexual reproduction by *meiosis* produces cells that contain half the genetic information of the parent cell. Although mitosis and meiosis have many similarities, they also have some very important differences that are essential to recognize. In this chapter, I present the details of cell division and show you how to recognize the phases of mitosis and meiosis.

Talking 'bout the Generations

Cells multiply for the following important reasons:

✔ To make copies of cells for growth

✔ To make copies of cells for repair

✔ To carry on the species

Some organisms, like bacteria, make new generations through *asexual reproduction* by making exact copies of their cells. Other organisms, like humans, make new generations through *sexual reproduction* by producing special cells, called *gametes,* that contain half the amount of genetic information that a new organism needs, like when your mom and dad made the egg and sperm that joined to form you.

When cells reproduce, they make copies of all their parts, including their DNA, and then divide themselves up to make new cells. After a cell copies the DNA, it divides either once or twice depending on whether it's reproducing sexually or asexually:

✔ Cells participating in asexual reproduction divide once through a process called *mitosis,* creating cells that are exact duplicates of the original cell.

- ✔ Cells participating in sexual reproduction divide twice through a process called *meiosis,* creating cells that have half the genetic material of the original cell.

Q. The largest organism ever found on the planet is a colony of the honey mushroom (*Armillaria ostoyea*), growing in Oregon, that spread over 2,200 acres and weighed 605 tons. Like many mushrooms, the honey mushroom grows by making exact copies of its cells to build fine threads that spread through soil to form a huge network called a *mycelium.* Periodically, mushroom colonies enter sexual reproduction, which is when they build the mushroom-shaped structures that you're familiar with. In the cap of the mushroom, cells divide to produce reproductive cells called *basidiospores* that have half of the genetic information of the parent.

When the mushroom is growing as threads through the soil, what type of cell division is it using (mitosis or meiosis)? If a single basidiospore contains 0.065 picograms of DNA (a *picogram* is a trillionth of a gram), how much DNA would you expect to find in a cell of the mushroom?

A. When the mushroom is growing through the soil, it's producing identical cells, so it must be dividing by mitosis. Basidiospores are made for sexual reproduction, and they have half the genetic material of the parent. Because $0.065 \times 2 = 0.13$, a cell of the mushroom itself should have 0.13 picograms of DNA.

Questions 1 and 2 refer to the following scenario:

The starlet sea anemone is an animal that lives in salt marshes along the coastlines of North America and Great Britain. It burrows into the mud and uses feeding tentacles to catch prey in the water. Like most anemones, this animal can reproduce both sexually and asexually.

1. A starlet sea anemone is injured and must make new cells to repair one of its tentacles. Which process would it use?

   a. Mitosis

   b. Meiosis

2. You remove a single cell from the body of the anemone and weigh its DNA. The total DNA weighs 0.46 picograms (a *picogram* is a trillionth of a gram). If a cell from the sexual organs of the anemone divided by meiosis to make sperm or egg cells, how much would the DNA from a single sperm or egg cell weigh?

   a. 0.46 picograms

   b. 0.23 picograms

   c. 0.115 picograms

   d. 0.92 picograms

Duplicating and Dividing Your DNA

Cells that are about to divide use *DNA replication* to copy their genetic material. DNA molecules form a double helix that looks like a twisted ladder, with the nucleotide bases forming the "rungs" of the ladder. (See Chapter 2 for a picture of a DNA molecule.) In cells, each double helix coils around proteins, forming structures called *chromosomes.* Different types of organisms have unique numbers of chromosomes. Whatever number of chromosomes it has, a cell has to duplicate each one before cell division can occur.

During DNA replication, cells separate the two original DNA strands of each double helix and use each one as a pattern, or template, for the building of a new strand. Thus, at the end of replication, each new double helix is made of one old strand and one new strand. Scientists call this *semiconservative* replication because half of the old DNA is saved, or conserved, for each new double helix that's made (but note that the word *semiconservative* may not apply to your political science class).

Q. Figure 5-1 shows part of the DNA double helix from a single chromosome cell. Original DNA and new DNA are represented by two different shades. If the DNA molecule is copied by semiconservative replication, which of the four possible products will result?

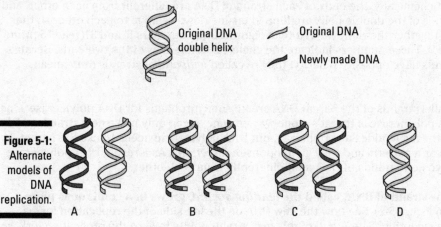

Figure 5-1:
Alternate
models of
DNA
replication.

A. The answer is C, because that answer shows two DNA molecules, each of which has one original strand and one new strand.

Several enzymes work together to make sure the DNA gets copied correctly. The enzyme that does the actual copying is called *DNA polymerase* because it's an enzyme (so its name ends in *-ase*) that makes *polymers* (chains) of DNA. See Figure 5-2 for a drawing that shows DNA polymerase and its partner enzymes hard at work copying DNA.

The basic steps of DNA replication are as follows:

1. **The two parental DNA strands separate so that the rungs of the ladder are split apart with one nucleotide on one side and one nucleotide on the other.**

   The entire DNA strand doesn't unzip all at one time, however. Only part of the original DNA strand opens up at one time. The partly open/partly closed area where the replication actively happens is called the *replication fork* (this is the Y-shaped area in Figure 5-2).

2. **DNA polymerase reads the DNA code on the parental strands and builds new partner strands that are complementary to the original strands.**

   To build complementary strands, DNA polymerase follows the *base-pairing rules* for the DNA nucleotides: adenine (A) always pairs with thymine (T), and cytosine (C) always pairs with guanine (G) (see Chapter 2 for more on nucleotides).

Several enzymes help DNA polymerase with the process of DNA replication (you can see them in Figure 5-2):

- **Helicase** separates the original parental strands to open the DNA.

- **Primase** puts down short pieces of RNA, called *primers,* that are complementary to the parental DNA. DNA polymerase needs these primers in order to get started copying the DNA.

- **DNA polymerase I** removes the RNA primers and replaces them with DNA, so it's slightly different from the DNA polymerase that makes most of the new DNA (that enzyme is officially called *DNA polymerase III*).

- **DNA ligase** forms covalent bonds in the backbone of the new DNA molecules to seal up the small breaks created by the starting and stopping of new strands.

The parental strands of the double helix are oriented to each other in *opposite polarity,* meaning that chemically, the ends of each strand of DNA are different from each other, and the two strands of the double helix are flipped upside down relative to each other so that the bases fit together the right way. Note in Figure 5-2 the numbers 5' and 3' (read "5 prime" and "3 prime"). These numbers indicate the chemical differences of the two ends. Because the two strands have opposite polarity, they're called *antiparallel* strands (*anti-* means "opposite").

The antiparallel strands of the parent DNA create some problems for DNA polymerase. One quirk of DNA polymerase is that it's a one-way enzyme; it can only make new strands of DNA by lining up the nucleotides a certain way. But DNA polymerase needs to use the parent DNA strands as a pattern, and they go in opposite directions. As a result, DNA polymerase makes the two new strands of DNA a bit differently from each other:

- **One new strand of DNA, called the *leading strand,* grows in a continuous piece** (refer to Figure 5-2). See how the new DNA on the left side of the replication fork is growing smoothly? The 3' end of this new strand points toward the replication fork, so after DNA polymerase starts building the new strand, it can just keep going.

- **One new strand of DNA, called the *lagging strand,* grows in fragments.** Look at Figure 5-2 again. Notice how the right side of the replication fork looks a little messier? That's because the replication process doesn't occur smoothly over there. The 3' end of this new strand points away from the fork. DNA polymerase starts making a piece of this new strand but has to move away from the fork to do so (because it can only work in one direction). DNA polymerase can't go too far from the rest of the enzymes that are working at the fork, however, so it has to keep backing up toward the fork and starting over. As a result, the lagging strand is made in lots of little pieces called *Okazaki fragments.* After DNA polymerase is done making the fragments, the enzyme DNA ligase comes along and forms covalent bonds between all the pieces to make one continuous new strand of complementary DNA.

3. Before scientists figured out that DNA replication is semiconservative, they had other ideas about how the process might work. One of these hypotheses was called the *conservative model* because it proposed that, after the DNA was copied, all the original DNA would stay together and all the new DNA would get together. Refer back at Figure 5-1: Which of the choices in that figure accurately represents the conservative model of DNA replication?

4.–8. Match the following job descriptions to the enzyme that performs the task.

    a. Finishes the job of DNA replication by forming covalent bonds in the sugar-phosphate backbone between fragments.

    b. Starts the process of DNA replication by separating the parental strands from each other.

    c. Makes new strands of DNA that are complementary to the original parental strands.

    d. Lays down short pieces of RNA so that DNA polymerase can start building.

    e. Removes pieces of RNA and replaces them with DNA.

4. Helicase

5. Primase

6. DNA polymerase (III)

7. DNA polymerase (I)

8. DNA ligase

9.–16. Use the terms that follow to identify the enzymes and components of the replication fork shown in Figure 5-2.

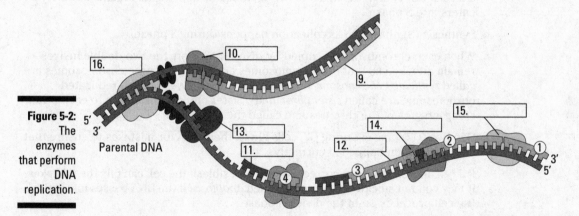

Figure 5-2: The enzymes that perform DNA replication.

    a. DNA polymerase (III)

    b. DNA ligase

    c. Primase

    d. Helicase

    e. DNA polymerase (I)

    f. Leading strand

    g. Lagging strand

    h. Primer

17. If a cell is going to copy a DNA molecule by DNA replication, which of the following enzymes must do its job before any of the others can do theirs?

a. DNA polymerase (DNA polymerase III)

b. DNA polymerase I

c. Primase

d. DNA ligase

Riding the Cell Cycle

DNA replication occurs just before cells are about to divide, as part of a sequence of events that scientists call the *cell cycle*. The cell cycle represents the life of a cell as it alternates between dividing and nondividing states.

1. **The nondividing part of the cell cycle is called *interphase*.**

   During interphase, cells go about their regular business. Interphase has three subphases:

   a. Gap one (G1) phase: During G1, cells function normally.

      If a cell in G1 receives a signal to divide, it begins copying its organelles and then enters into S phase.

   b. Synthesis (S) phase: DNA replication happens during S phase.

      When each chromosome is copied by DNA replication, the two double helixes remain attached so that the chromosomes are doubled. These chromosomes are called *replicated chromosomes*. The two identical halves of each replicated chromosome are called *sister chromatids*. Sister chromatids attach to each other along a region of the chromosome called the *centromere*.

   c. Gap two (G2) phase: During G2, cells check their DNA for mistakes to ensure that DNA replication happened correctly.

      If a cell finds mistakes, it corrects them if possible. If the cell can't fix the mistakes, it may commit suicide by a process called *apoptosis*. If the DNA passes inspection, the cell proceeds on to the division phase.

2. **Cells that receive a signal to divide enter a division process, which is either mitosis or meiosis.**

   Part of mitosis and meiosis is making sure that new cells get the proper number and right kinds of chromosomes. This can be tricky in cells that have multiple chromosomes, so cells must carefully choreograph both types of cell division.

The nuclear membrane is intact throughout interphase, as you can see in Figure 5-3. The DNA is loosely spread out, and you can't see individual chromosomes.

Questions 18 through 20 refer to the following scenario:

Assume that you remove a single cell that has just entered G1 from the body of a starlet sea anemone. You isolate the DNA from the single cell and weigh it, finding that the DNA weighs 0.46 picograms.

18.–20. Choose from the following to indicate the weight of the DNA for the same cell at each point in the cell cycle.

   a. 0.46 picograms

   b. 0.23 picograms

   c. 0.115 picograms

   d. 0.92 picograms

18. At the end of G1, how much would the DNA from a single cell weigh?

19. At the end of S phase, how much would the DNA from a single cell weigh?

20. At the end of G2, how much would the DNA from a single cell weigh?

Marching Through Mitosis

During mitosis and meiosis, the cytoskeletal proteins of the cell form a network of *spindle fibers* called the *mitotic spindle* that sorts the chromosomes (see the thin curving lines drawn in the cells in Figure 5-3). In animal cells, small bundles of cytoskeletal proteins form cylindrical structures called *centrioles* that are visible in the organizing centers for the mitotic spindle (called, appropriately enough, *microtubule organizing centers* or MTOC).

Although the cell cycle is a continuous process, with one stage flowing into another, scientists divide the events of mitosis into four phases based on the major events in each stage:

   1. *Prophase:* **The cell's replicated chromosomes get ready to be moved around by coiling themselves up into tight little packages.**

   As the chromosomes coil up, or *condense,* they become visible to the eye when viewed through a microscope. Also during prophase:

   • The nuclear membrane breaks down.

   • The mitotic spindle forms and attaches to the chromosomes.

   • The nucleoli break down and become invisible.

   2. *Metaphase:* **The mitotic spindle tugs the replicated chromosomes until they're all lined up in the middle of the cell.**

3. *Anaphase:* The replicated chromosomes separate so that the two sister chromatids from each replicated chromosome go to opposite sides.

4. *Telophase:* The cell gets ready to divide into two by forming new nuclear membranes around the separated sets of chromosomes.

The events of telophase are essentially the reverse of prophase:

   • The chromosomes uncoil and spread out through the nucleus.

   • New nuclear membranes form around the two sets of chromosomes.

   • The mitotic spindle breaks down.

   • The nucleoli reform and become visible again.

The last order of cell-division business is to give the new daughter nuclei their own cells through a process called *cytokinesis* that divides up the cytoplasm of the two cells. Cytokinesis occurs differently in animal and plant cells (see Figure 5-5):

✔ In animal cells, cytokinesis begins with an indentation, called a *cleavage furrow,* in the cell's center. Cytoskeletal proteins act like a belt around the cell, contracting down and squeezing the cell in two.

✔ In plant cells, a new cell wall forms at the cell's center. Little bags made of membrane, called *vesicles,* carry wall material to the center of the cell (see Chapter 3 for more on vesicles). The vesicles fuse together, and their membranes form the plasma membranes of the new cells. The wall material gets dumped between the new membranes, forming the plant's cell walls.

After cytokinesis is complete, the new cells move immediately into the G1 stage of interphase.

Q. A cell that's in metaphase is going to complete mitosis, and then its descendants will divide again. What are all the phases of the cell cycle that the cells will go through until they return again to metaphase?

A. After metaphase, the cells move into anaphase and then telophase. Cell division finishes with cytokinesis, putting the descendants into interphase. From G1 of interphase, the cells enter S phase, then G2, and then they begin division with prophase before returning again to metaphase.

Try these questions to test your understanding of mitosis:

21.–24. Use the following terms to label the phases of mitosis in Figure 5-3.

   a. Telophase

   b. Metaphase

   c. Anaphase

   d. Prophase

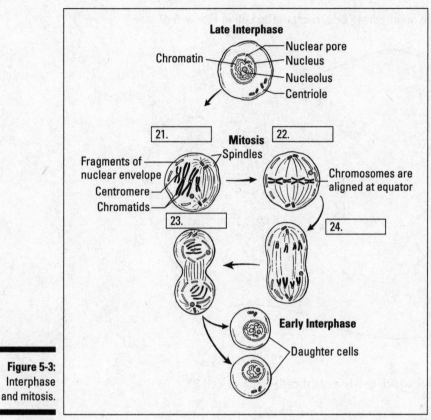

Figure 5-3:
Interphase
and mitosis.

Illustration by Kathryn Born, M.A

25.–27. Use the terms that follow to identify the correct phase of the cell cycle for each event.

   a. Interphase

   b. Prophase

   c. Anaphase

   d. Telophase

   e. Metaphase

25. During which phase of the cell cycle would the nuclear membrane be visible for the entire length of the phase?

26. During which phase of mitosis do sister chromatids separate?

27. What phase of mitosis is occurring in the cell in Figure 5-4?

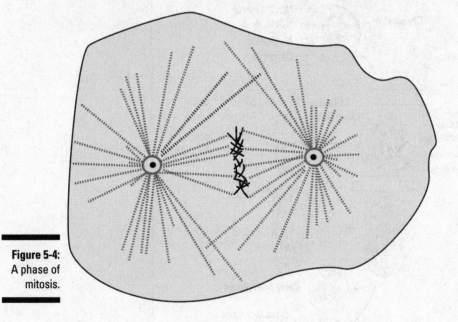

Figure 5-4:
A phase of
mitosis.

28. In Figure 5-5, which cell is a plant cell, cell A or cell B?

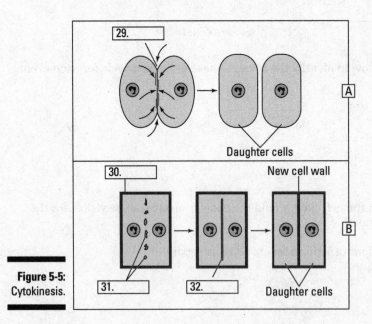

Figure 5-5:
Cytokinesis.

29.–32. Use the terms that follow to label the structures of cytokinesis in Figure 5-5.

     a. Vesicles containing wall material

     b. Contractile ring

     c. Cell plate

     d. Cell wall

Getting Ready for Sexual Reproduction

The biggest difference between *meiosis* and mitosis is that meiosis produces cells with only half of their parent cells' *chromosomes*. Meiosis is the type of cell division that separates chromosomes so gametes receive one of each type of chromosome. Human body cells have 46 chromosomes, 2 each of 23 different kinds. The 23 pairs of chromosomes can be sorted by their physical similarities and lined up to form a chromosome map called a *karyotype* (see Figure 5-6). The two matched chromosomes in each pair are called *homologous chromosomes* (*homo*- means "same," so these are chromosomes that have the same kinds of genetic information). In each pair of your homologous chromosomes, one chromosome came from your mom, and one came from your dad, so you have two copies of every gene (with the exception of genes on the X and Y chromosomes if you're male).

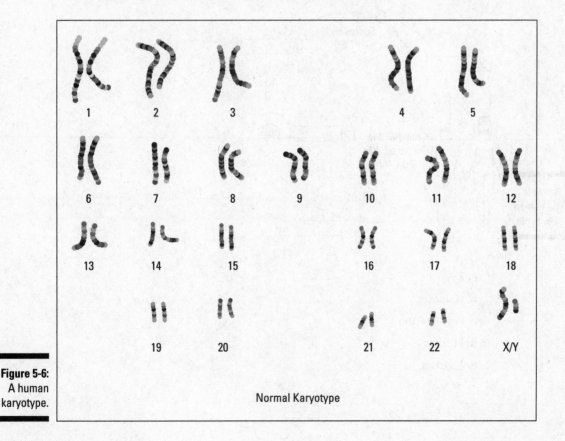

Figure 5-6:
A human
karyotype.

Normal Karyotype

Human gametes have just 23 chromosomes. Through *fertilization* (refer to Figure 5-7), a sperm and an egg join to create a cell called a *zygote,* returning the chromosome number to 46. The zygote divides by mitosis to make all the cells of the human body.

In humans, meiosis separates the 23 pairs of chromosomes so that each cell receives just *one of each pair.* Consequently, gametes have what's known as a *haploid* number of chromosomes, or a single set. When the two gametes unite, they combine their chromosomes to reach the full complement of 46 chromosomes in a normal *diploid* cell (one with a double set of chromosomes, or two of each type).

33. Circle one pair of homologous chromosomes on Figure 5-6.

34. Using two different colored pencils or highlighters, mark Figure 5-6 to show the results of meiosis. Use one color to mark one set of chromosomes that could be given to a gamete. Then use the other color to mark the remaining set of chromosomes.

35.–39. Use the terms that follow to label the human life cycle in Figure 5-7.

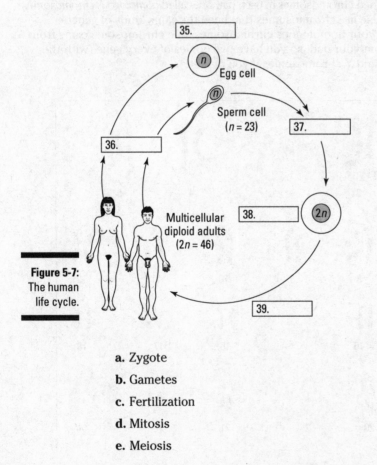

35.

n

Egg cell

n

Sperm cell
(*n* = 23)

37.

36.

38.

2*n*

Multicellular
diploid adults
(2*n* = 46)

39.

Figure 5-7:
The human
life cycle.

a. Zygote

b. Gametes

c. Fertilization

d. Mitosis

e. Meiosis

Making Gametes by Meiosis

Two cell divisions occur in meiosis, and the two halves of meiosis are called *meiosis I* and *meiosis II* (see Figure 5-8).

> ✔ During meiosis I, homologous chromosomes are paired up and then separated into two daughter cells. Each daughter cell receives one of each chromosome pair, but the chromosomes are still replicated. (You can see in Figure 5-8 that the chromosomes still look like *X*'s after meiosis I.)

> ✔ During meiosis II, the replicated chromosomes send one sister chromatid from each replicated chromosome to new daughter cells. After meiosis II, four daughter cells each have one of each chromosome pair, and the chromosomes are no longer replicated. (Notice in Figure 5-8 how the four daughter cells don't have sister chromatids.)

Normal Meiosis

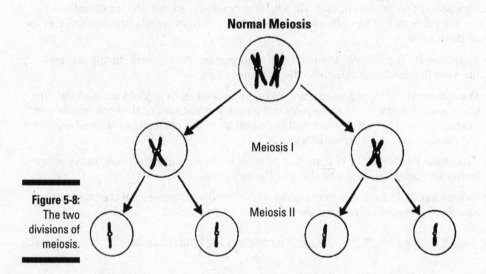

Meiosis I

Meiosis II

Figure 5-8:
The two divisions of meiosis.

The events that occur during meiosis have many similarities to those in mitosis. One of the most important differences in meiosis is that homologous chromosomes pair up during prophase I. (Homologous chromosomes don't pair at all during mitosis.) The phases of meiosis are as follows:

> ✔ **Meiosis I:** The cell goes from diploid to haploid as the chromosomes in each pair of homologs separate from each other:
>
> • **Prophase I:** During this phase, the cell's nuclear membrane breaks down, the chromatids coil to form visible chromosomes, the nucleoli break down and disappear, and the spindles form and attach to the chromosomes. But that's not all. Prophase I is when something that's absolutely critical to the successful separation of homologous chromosomes occurs: synapsis.
>
>   *Synapsis* happens when the two chromosomes of each homologous pair find each other and stick together. At this point, the two homologous chromosomes can swap equal amounts of DNA in an event called *crossing-over*.
>
> • **Metaphase I:** This is when the pairs of homologous chromosomes line up in the cell's center.

- **Anaphase I:** During this phase, the two members of each homologous pair go to opposite sides of the cell, and the chromosome number is officially reduced from diploid to haploid.

- **Telophase:** This is when the cell takes a step back (or forward, depending on your perspective) to an interphase-like condition by reversing the events of prophase I. Specifically, the nuclear membrane reforms, the chromosomes uncoil and spread throughout the nucleus, the nucleoli reform, and the spindles break down.

Crossing-over between homologous chromosomes during prophase I increases the genetic variability among gametes produced by the same organism. Every time meiosis occurs, crossing-over can happen a little differently, shuffling the genetic deck as gametes are made. This is one of the reasons that siblings can be so different from each other.

✔ **Meiosis II:** Sister chromatids of each replicated chromosome move to opposite sides of the cell. Both daughter cells produced by meiosis I divide again to produce a total of four gametes. The phases of meiosis II look very similar to the phases of mitosis I, with one big exception: The cells start out with half the number of chromosomes as the original parent cell.

- **Prophase II:** The nuclear membrane disintegrates, the nucleoli disappear, and the spindles form and attach to the chromosomes.

- **Metaphase II:** Nothing too exciting here, folks. Just as in any old metaphase, the chromosomes line up at the equatorial plane. But remember that the number of chromosomes that line up are half the number of the original parent cell (and half the number you'd see in mitosis).

- **Anaphase II:** The sister chromatids of each replicated chromosome move away from each other to opposite sides of the cell.

- **Telophase II:** The nuclear membrane and nucleoli reappear, the chromosomes uncoil, and the spindles disappear.

After meiosis II, it's time for cytokinesis, which creates four haploid cells.

40. Use a pencil to label the pair of homologous chromosomes in Figure 5-8. Draw a square around a replicated chromosome and label the two sister chromatids.

41.–51. Use the terms that follow to identify the phases of meiosis and important cellular structures depicted in Figure 5-9.

    a. Prophase I

    b. Metaphase I

    c. Anaphase I

    d. Telophase I

    e. Nucleus

    f. Spindle fiber

    g. Centromere

    h. Prophase II

    i. Metaphase II

    j. Anaphase II

    k. Telophase II

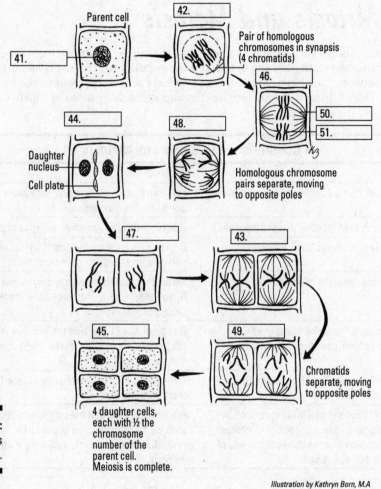

Parent cell

42.

Pair of homologous chromosomes in synapsis (4 chromatids)

41.

46.

50.

51.

44.

48.

Daughter nucleus

Cell plate

Homologous chromosome pairs separate, moving to opposite poles

47.

43.

45.

49.

4 daughter cells, each with ½ the chromosome number of the parent cell. Meiosis is complete.

Chromatids separate, moving to opposite poles

Figure 5-9:
The events of meiosis.

Illustration by Kathryn Born, M.A

52. During prophase II of meiosis, are cells diploid or haploid?

    a. Diploid

    b. Haploid

53. During which phase of meiosis does crossing-over occur?

    a. Prophase I

    b. Prophase II

    c. Metaphase I

    d. Anaphase I

54. During which phase of meiosis do sister chromatids separate?

    a. Metaphase I

    b. Metaphase II

    c. Anaphase I

    d. Anaphase II

Contrasting Mitosis and Meiosis

The phases of meiosis are very similar to the phases of mitosis; they even have the same names, which can make distinguishing between the two rather tough. Just remember that the key difference between the phases of mitosis and meiosis is what happens to the number of chromosomes. Table 5-1 can help you sort out the important differences at a glance.

Table 5-1	A Comparison of Mitosis and Meiosis
Mitosis	*Meiosis*
One division is all that's necessary to complete the process.	Two separate divisions are necessary to complete the process.
Chromosomes don't get together in pairs (synapse).	Homologous chromosomes synapse in prophase I.
Homologous chromosomes don't cross over.	Crossing-over is an important part of meiosis and one that leads to genetic variation.
Sister chromatids separate in anaphase.	Sister chromatids separate only in anaphase II, not anaphase I. (Homologous chromosomes separate in anaphase I.)
Daughter cells have the same number of chromosomes as their parent cells, meaning they're diploid.	Daughter cells have half the number of chromosomes as their parent cells, meaning they're haploid.
Daughter cells have genetic information that's identical to that of their parent cells.	Daughter cells are genetically different from their parent cells.
The function of mitosis is asexual reproduction in some organisms. In many organisms, mitosis functions as a means of growth, replacement of dead cells, and damage repair.	Meiosis creates gametes or spores, the first step in the reproductive process for sexually reproducing organisms, including plants and animals.

Q. A cell that has a diploid number of four chromosomes is undergoing cell division, as shown in Figure 5-10. Based on the information in the figure, which phase or phases of mitosis or meiosis are possible?

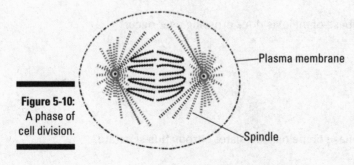

Figure 5-10: A phase of cell division.

Plasma membrane

Spindle

A. The figure shows sister chromatids from four chromosomes moving away from each other. The diploid number of chromosomes for this cell is four, so the cell in Figure 5-10 has the full number of chromosomes. This means you can rule out meiosis II entirely (cells in meiosis II would already have half the diploid number of chromosomes). If a diploid cell with four chromosomes were in meiosis I, then you'd expect that the four chromosomes would have formed two pairs of homologous chromosomes and that homologs would separate during anaphase I. The chromosomes in the picture aren't paired, and the separation involves sister chromatids, not homologous chromosomes, so you can also rule out meiosis I.

After eliminating meiosis, mitosis is left. If a diploid cell with four chromosomes went through mitosis, the four replicated chromosomes would line up in the middle of the cell, and their sister chromatids would separate during anaphase. That's exactly what's shown in Figure 5-10. So the answer is anaphase of mitosis.

55.–57. For each cell, identify the correct appearance of the chromosomes by choosing from the following list:

  a. Eight replicated chromosomes lined up individually (not in pairs)

  b. Eight replicated chromosomes lined up in four pairs

  c. Four replicated chromosomes lined up individually (not in pairs)

  d. Eight unreplicated chromosomes lined up in four pairs

55. A cell with a diploid number of eight chromosomes undergoes mitosis. If you looked at this cell under a microscope during metaphase, what would you see?

56. A cell with a diploid number of eight chromosomes undergoes meiosis. If you looked at this cell under a microscope during metaphase I, what would you see?

57. A cell with a diploid number of eight chromosomes undergoes meiosis. If you looked at this cell under a microscope during metaphase II, what would you see?

58. Take another look at Figure 5-10. Imagine that it's a different cell than the one described in the previous example. Instead, imagine that you're looking at the division of a cell with a diploid number of *eight* chromosomes. If Figure 5-10 shows a cell with a diploid number of eight chromosomes, which phase of mitosis or meiosis is occurring in this cell?

  a. Anaphase of mitosis

  b. Anaphase I of meiosis

  c. Anaphase II of meiosis

Answers to Questions on Cell Division

The following are answers to the practice questions presented in this chapter.

1 The answer is **a. Mitosis.**

2 The answer is **b. 0.23 picograms.**

Meiosis reduces the amount of genetic information by half.

3 The answer is **A.**

A shows two copies of the chromosome: one that's all original DNA and one that's all new DNA.

4 The answer is **b. Starts the process of DNA replication by separating the parental strands from each other.**

5 The answer is **d. Lays down short pieces of RNA so that DNA polymerase can start building.**

6 The answer is **c. Makes new strands of DNA that are complementary to the original parental strands.**

7 The answer is **e. Removes pieces of RNA and replaces them with DNA.**

8 The answer is **a. Finishes the job of DNA replication by forming covalent bonds in the sugar-phosphate backbone between fragments.**

9 – 16 The following is how Figure 5-2 should be labeled:

9 **f. Leading strand;** 10 **a. DNA polymerase (III);** 11 **h. Primer;** 12 **g. Lagging strand;** 13 **c. Primase;** 14 **e. DNA polymerase (I);** 15 **b. DNA ligase;** 16 **d. Helicase**

17 The answer is **c. Primase.**

18 The answer is **a. 0.46 picograms.**

At the end of G1, the DNA hasn't yet been copied, so the amount remains the same.

19 The answer is **d. 0.92 picograms.**

DNA is copied during S phase, so by the end of S phase, the weight of DNA will have doubled.

20 The answer is **d. 0.92 picograms.**

During G2, cells check the doubled DNA, and separation of the DNA hasn't yet occurred, so the amount of DNA would be the same as in S phase.

21 – 24 The following is how Figure 5-3 should be labeled:

21 **d. Prophase;** 22 **b. Metaphase;** 23 **a. Telophase;** 24 **c. Anaphase**

25 The answer is **a. Interphase.**

Nuclear membranes are visible throughout interphase, and they break down and reform during prophase and telophase.

26 The answer is **c. Anaphase.**

27 The answer is **e. Metaphase.**

The chromosomes are lined up in the middle of the cell, which only occurs during metaphase.

28 The answer is **cell B.**

You can tell that the bottom cell is a plant cell because it looks rigid and boxy. Cell division is occurring as a line forms down the center of the dividing cells. In comparison, the animal cell in A looks rounded and flexible because it's being pinched into two.

29–32 The following is how Figure 5-5 should be labeled:

29 **b. Contractile ring;** 30 **d. Cell wall;** 31 **a. Vesicles containing wall material;** 32 **c. Cell plate**

33 You could circle any of the numbered pairs.

34 Using one color, you should have marked one of every pair (it doesn't matter which one). Using the other color, you should have marked the remaining member of every pair. All together, you should have highlighted 23 chromosomes with one color and 23 with the other color. Each numbered pair should have one chromosome highlighted with each other.

35–39 The following is how Figure 5-7 should be labeled:

35 **b. Gametes;** 36 **e. Meiosis;** 37 **c. Fertilization;** 38 **a. Zygote;** 39 **d. Mitosis**

40 Only the very first cell shows a pair of homologous chromosomes, so you should have circled that pair. You could have drawn a square around any of the chromosomes that looks like an X. You'd label each half of the X as a sister chromatid.

41–51 The following is how Figure 5-9 should be labeled:

41 **e. Nucleus;** 42 **a. Prophase I;** 43 **i. Metaphase II;** 44 **d. Telophase I;** 45 **k. Telophase II;** 46 **b. Metaphase I;** 47 **h. Prophase II;** 48 **c. Anaphase I;** 49 **j. Anaphase II;** 50 **f. Spindle fiber;** 51 **g. Centromere**

52 The answer is **b. Haploid.**

Homologous pairs of chromosomes separate during meiosis I. After that occurs, cells are haploid because the genetic information that originally came from the parents has separated. Even though the chromosomes are still replicated, that doesn't count as making cells diploid because sister chromatids are identical.

53 The answer is **a. Prophase I.**

54 The answer is **d. Anaphase II.**

55 The answer is **a. Eight replicated chromosomes lined up individually (not in pairs).**

Chromosomes are replicated in S phase and haven't yet separated by metaphase, so they're still replicated (sister chromatids present). The diploid number of chromosomes is eight and no separation has occurred, so eight chromosomes are still present. During mitosis, homologs don't pair, so the chromosomes are lined up individually.

56 The answer is **b. Eight replicated chromosomes lined up in four pairs.**

A cell that has a diploid number of eight chromosomes has four pairs of chromosomes (just like humans that have a diploid number of 46 chromosomes have 23 pairs of chromosomes). During meiosis I, the pairs of homologous chromosomes synapse together and line up together.

57 The answer is **c. Four replicated chromosomes lined up individually (not in pairs).**

The four pairs of homologous chromosomes in this cell would have already separated during meiosis I. So by metaphase II, chromosomes would line up individually, and each cell would have half as many.

58 The answer is **c. Anaphase II of meiosis.**

The diploid number of chromosomes for the cell is eight, but the cell only shows four chromosomes in the act of separating. Therefore, the cell must be in meiosis II (half the chromosomes have already moved to another cell). The cell shows sister chromatids separating, which occurs during anaphase.

Chapter 6

Predicting Future Generations with Mendelian Genetics

In This Chapter

▶ Speaking the vernacular of genetics

▶ Understanding how genetic traits are inherited

▶ Using test crosses to determine the traits of progeny

▶ Analyzing human pedigrees

Genetics is the branch of biology that examines how parents pass on traits to their offspring. Genetics started more than 150 years ago, when a monk named Gregor Mendel conducted breeding experiments with pea plants that led him to discover the fundamental rules of inheritance. Although Mendel worked with peas, his ideas explain a lot about why you look and function the way you do.

In this chapter, I show you how Mendel figured out some of the most important rules of inheritance and give you a chance to practice your skills.

Getting Acquainted with the Lingo of Genetics

Geneticists use a vocabulary of their own that includes terms like *genes, alleles, dominant,* and *recessive.* To understand genetics and solve genetics problems, you need to memorize and be able to apply the basic terms that geneticists use:

✔ **Chromosome, genes, alleles, and locus:** A *chromosome* is a piece of DNA. *Genes* are individual blueprints written in the DNA code along a chromosome. *Alleles* are different forms of the same gene. Scientists call the place on a chromosome where a particular gene is located the *locus.*

✔ **Genotype and phenotype:** The genes, or DNA codes, that an organism has are its *genotype.* How an organism looks and functions as a result of those codes is its *phenotype.*

✔ **Heterozygous and homozygous:** If an organism has two different alleles for a particular gene, it's *heterozygous* for that gene. If an organism has two identical alleles for a gene, it's *homozygous* for that gene.

✔ **Dominant and recessive:** Sometimes when an organism is heterozygous, one allele completely hides the effect of the other allele. An allele that hides another allele is called *dominant,* while an allele that's hidden in the phenotype is called *recessive.*

1.–4. Use the following terms to fill in the blanks:

    a. Chromosomes

    b. Gene

    c. Locus

    d. Allele

1. Humans have 46 pieces of DNA called _____ in the nucleus of each cell.

2. Chromosome 1 in humans contains a blueprint, or _____, for a protein called Rh factor that determines whether a person has a positive or negative blood type.

3. Two forms of the gene for Rh factor exist in the human population. The _____ of the gene that contains the code for a functional protein contributes to positive blood types.

4. Each person has two copies of the gene for Rh factor, one from his mother and one from his father. The gene for Rh factor can be found in the same place, or _____, on chromosome 1 from each parent.

5.–8. Use the following terms to fill in the blanks.

    a. Genotype

    b. Phenotype

    c. Heterozygous

    d. Homozygous

    e. Dominant

    f. Recessive

5. Whether a person has a positive or negative blood type represents her _____.

6. A person who has one allele for a functional Rh factor and one allele for a nonfunctional Rh factor is _____ for the gene for Rh factor.

7. A person who's heterozygous for the gene for Rh factor has a positive blood type, even though one of his alleles is for nonfunctional Rh protein. Thus, the allele for functional Rh factor is _____ to the allele for nonfunctional protein.

8. If a person is homozygous for the nonfunctional allele for Rh factor, this represents her _____ for the trait.

Discovering the Laws of Inheritance

Every individual has a unique combination of traits that result from the interaction between their genetic material and their environment. Through *meiosis*, parents make copies of their chromosomes, passing one of each kind of chromosome to their offspring via *gametes* (egg or sperm; for more on meiosis and gametes, head to Chapter 5). Along each chromosome are blueprints, called *genes,* that tell cells how to build the molecules that determine an organism's structure and function. The interaction between the genes from each parent determines the *heritable traits* of the offspring.

To figure out how traits are inherited, geneticists perform carefully planned breeding experiments. Back in the 1850s, the Austrian monk Gregor Mendel founded the entire science of genetics based on his breeding experiments with garden peas, before scientists had even discovered chromosomes or meiosis. His discoveries have stood the test of time and remain valid today.

Mendel studied many characteristics of garden peas, including flower color, pea shape, pea color, and plant height. For each trait, he did a controlled mating, called a *genetic cross,* to determine how the trait was inherited:

✔ **The organisms that are mated first in a cross are called the *parental generation* (P1).** In order to know for sure what genetic message a parent gives to its offspring, parentals are always pure breeding. *Pure breeding* organisms always reproduce the same version of a trait in their offspring.

For example, when Mendel studied pea plant height, he used pure breeding tall plants and pure breeding short plants as the parentals in his first cross. When mated to themselves, pure breeding tall plants always produced tall offspring and pure breeding short plants always produced short offspring. Because they bred true, Mendel knew that the tall parentals only had messages for tallness in their chromosomes, and the short parentals only had messages for shortness in their chromosomes. (The messages that Mendel was thinking about are what scientists today call *alleles.*)

✔ **The first offspring generation in a cross is called the *F1 generation.*** F1 stands for "first filial"; *filial* means offspring.

Mendel's mating between tall and short parentals produced all tall plants in the F1. Mendel knew that the short parental must have given an allele for shortness to the offspring (because they were true breeding!), so he knew that the allele for tallness must be hiding the allele for shortness.

✔ **When members of the F1 generation are mated to each other, the next offspring generation is called the *F2 generation.*** F2 stands for "second filial."

When Mendel mated his tall F1 plants with each other, he got both tall and short pea plants in the F2 generation. Specifically, for every short plant in the F2, he got three tall plants in the same F2 — in other words, a 3:1 ratio of tall plants to short plants.

From his results, Mendel figured out several things that are fundamental to the science of genetics. (Note that I've expressed these things using modern language like genes and alleles, which isn't how Mendel referred to them. In Mendel's time, no one even knew that chromosomes existed, but he still managed to figure out the fundamental rules of inheritance!)

- ✔ **Every organism has two copies of each gene, one from mom and one from dad.** In other words, these organisms are *diploid*. (Refer to Chapter 5 for a discussion of haploid and diploid.)

- ✔ **Different forms of genes, called alleles, exist in the population.** For example, for the plant height gene, peas can have alleles for tallness or alleles for shortness.

- ✔ **One allele can hide another.** In Mendel's experiment, the allele for tallness hid the allele for shortness in the F1 generation. In modern language, geneticists say that the allele for tallness is *dominant* to the allele for shortness. The allele for shortness is *recessive* to the allele for tallness.

- ✔ **Parents give just one of their two copies (alleles) of a gene to their offspring.** And which copy they give is chosen randomly.

Sexually reproducing organisms have two copies of every gene, but they give only one copy of each gene to their offspring. Mendel said that this is because the two copies of genes *segregate* (separate from each other) when the organisms reproduce. Scientists now call this idea *Mendel's law of segregation*.

Modern geneticists use tools called *Punnett squares* to demonstrate the numbers and types of combinations of alleles that are possible during a genetic cross.

In a Punnett square, the possible allele combinations in the parent's gametes are written along the square's edge, and then all the possible combinations of gametes that could occur in offspring are written inside the square. Typically, biologists use a single letter of the alphabet to represent a single gene, with uppercase letters for dominant alleles and lowercase letters for recessive alleles.

Q. Draw two Punnett squares to illustrate the genetic crosses in Mendel's experiment with tall and short peas. One square should represent the first cross between the parentals. The other square should represent the cross between the plants of the F1 generation. Use an uppercase *T* to represent the allele for tallness, and a lowercase *t* to represent the allele for shortness.

A. The square on the left in Figure 6-1 represents the cross between the parentals. The tall parental was pure breeding for tallness, so it could give only tall alleles. It gave one copy of this allele to each gamete, so the alleles for the gametes are written along the left edge of the Punnett square. The short parental was pure breeding for shortness, so it could give only short alleles. It also gave one copy of its allele to each gamete, so the alleles for the gametes are written along the top of the square. All the F1 offspring, represented inside the square, had the genotype Tt.

The square on the right in Figure 6-1 represents the cross between F1 plants. The possible gametes from the F1 plants in this cross are written on the square's edges. The F1 plants are all Tt, so half the time they give a dominant allele (T) to the gamete, and the other half of the time they give a recessive allele (t). All the possible combinations of gametes that you'd see in the F2 form when the square is filled in. Notice that this time there's some variety! If two dominant alleles get together, the F2 offspring is TT. If one dominant and one recessive allele get together, the F2 offspring is Tt. When two recessives combine, the offspring is tt. The allele for tallness hides the allele for shortness, so three of these possible combinations (TT, Tt, and Tt) yield tall plants, while only one combination (tt) yields short plants. And again, that's exactly what Mendel observed — in his F2 generation, he saw a 3:1 ratio of tall plants to short plants.

Figure 6-1: Mendel's cross between tall and short pea plants.

	t	t
T	Tt	Tt
T	Tt	Tt

F1 Generation

	T	t
T	TT	Tt
t	Tt	tt

F2 Generation

All the possible combinations from the gametes of the parentals (imagine eggs meeting sperm) form when the square is filled in. In this case, the possible allele combinations will all be the same — T from the tall parental plus t from the short parental equals Tt for all F1 offspring. (Having all the same combination matches the observation in Mendel's cross that all the F1 were tall.)

REMEMBER

Crossing two organisms that are heterozygous for one trait results in a *monohybrid cross.* *Mono-* means "one," and *hybrid* means "something from two different sources," so a *monohybrid* is an organism that has two different alleles for one trait. The cross between F1 pea plants in Figure 6-1 is an example of a monohybrid cross.

EXAMPLE

Q. A pure breeding brown mouse is mated with a pure breeding white mouse. All the F1 offspring are brown. Which allele is dominant, brown or white? How do you know? What colors and ratios would you predict for mice in the F2 generation? Why?

A. The brown allele is dominant to the white allele. You know that the pure breeding brown mouse can only give messages for brown fur to the offspring, and the pure breeding white mouse can only give messages for white fur. So all the F1 offspring have one message for brown and one message for white, but they're only showing the brown allele. The brown allele hides the white allele, which means that brown is dominant to white.

If the brown allele is represented by B and the white allele is represented by b, then all F1 mice have Bb for their combination of alleles. (Note that I chose a B to represent the alleles because scientists typically use the first letter of the dominant allele.) When they're mated, the F1 parents give half of their gametes the B allele and the other half the b allele. If you draw a Punnett square, it will look exactly like the square on the right in Figure 6-1, except it has B's instead of T's. So the prediction for the F2 generation is the same — three dominant characteristics for every one recessive characteristic, or a 3:1 ratio of brown fur to white fur.

Now you try analyzing a cross to see whether you have the concept down pat.

9. A pure breeding tan fruit fly is mated with a pure breeding black fruit fly. All the offspring have tan bodies. Which body color is dominant? How do you know? What body colors and ratios would you predict for the F2 generation? Why? Draw a Punnett square to support your answer.

Figuring Out Genetic Puzzles

Geneticists don't always work from crosses of pure breeding organisms. Sometimes, they work with organisms that have an unknown genotype. In this case, one of the first things a geneticist does to figure out how a trait is inherited is to do a test cross.

A *test cross* is a mating between an organism of unknown genotype and an organism that's homozygous recessive for the trait being studied.

For example, suppose you have a brown mouse and you don't know its genotype. You know that brown fur is dominant to white fur in mice. The brown mouse could be homozygous dominant for brown fur (BB) or it could be heterozygous (Bb) — either way, it would still look brown! As a way to show what you know and what you don't know, you can write the mouse's genotype as B-, with the hyphen representing the unknown allele (B or b). To figure out the genotype of your mystery mouse, you do a test cross:

1. **You mate your mystery brown mouse (B-) with a pure breeding white mouse.**

   Because it's pure breeding, you know that the white mouse can only give the recessive allele.

2. **If your mystery mouse is homozygous dominant (BB), it's only able to give the brown allele to the offspring.**

   All offspring would get B from the brown mouse and b from the white mouse, so they'd all be Bb. If all the offspring of your test cross are brown, you know that the mystery mouse is BB.

3. **If your mystery mouse is heterozygous (Bb), half the time it gives the brown allele (B) to gametes, and half the time it gives the white allele (b).**

   The test cross mouse only gives b, so you'd predict that half the offspring would be Bb and the other half would be bb. In other words, half would be brown and half would be white. If you get this result, you know that your mystery mouse is Bb.

Q. In pea plants, the allele for purple flowers is dominant to the allele for white flowers. If a purple-flowered plant is crossed with a white-flowered plant and the result is 32 purple-flowered plants and 29 white-flowered plants, what's the genotype of the original purple-flowered plant? How do you know?

A. White flowers are recessive, so the white-flowered plant must be homozygous for the recessive allele (because if it had the purple allele, it would be purple, not white). Because the white-flowered plant is homozygous recessive, this is a test cross.

The white-flowered plant gives only the recessive allele (p) to offspring. If the purple-flowered plant were homozygous (PP), it would always give the dominant allele to offspring (P), and you'd predict that all offspring would be heterozygous (Pp) and purple. But this is not the case.

Instead, the offspring are about half purple and half white. If the purple plant were heterozygous (Pp), you'd expect that half the time it would give the dominant allele to offspring, producing offspring that are heterozygous and purple, and half the time it would give the recessive allele to offspring, producing offspring that are homozygous (pp) and white. This matches the outcome of the cross, so the original purple plant was heterozygous (Pp).

10.–12. As you answer Questions 10 through 12, bear in mind that, in pea plants, round seeds are dominant to wrinkled seeds.

10. If a heterozygous plant is crossed with a homozygous recessive plant, what types of seeds and ratios would you predict for their offspring?

11. If a plant with round seeds is crossed with a plant with wrinkled seeds, and all the offspring have round seeds, what's the genotype of the parent with round seeds? How do you know?

12. If two pea plants that are heterozygous for seed shape are crossed with each other, what types of seeds and ratios would you predict for their offspring?

Climbing the Branches of Your Family Tree

One of the reasons organisms like plants and fruit flies make good subjects for genetic studies is because you can control their mating. From a genetics standpoint, humans aren't nearly so cooperative. Also, humans don't produce as many offspring as plants and fruit flies do, and humans take a long time to grow up, so you have to wait for years to evaluate the appearance of traits. Consequently, when geneticists want to study human traits, they have to rely on families that already exist.

The first steps in understanding the inheritance of a human trait are to gather information on which people in a family have the trait and to record that information in a geneticist's version of a family tree, which is called a *pedigree*. The symbols in a pedigree (like the one shown in Figure 6-2) represent information about the family and the trait being studied:

✔ Squares indicate males; circles indicate females.

✔ A line between two symbols represents mating.

✔ A line drawn down from a mating indicates that the marriage produced a child. Children are arranged in birth order from left to right.

✔ A filled-in symbol indicates people who show the trait being studied; an open symbol is used for people who don't show the trait.

✔ A diagonal line through a symbol represents someone who's deceased.

✔ Roman numerals shown to the left of each row represent the different generations. Each person in a generation is assigned an Arabic number so that any person in a pedigree can be identified by the combination of his or her generation number and individual number. The deceased person in Figure 6-2c, for example, is identified as Individual VI-1.

By studying a pedigree, geneticists can often figure out whether a trait is caused by a dominant or recessive allele. The trait shown in Figure 6-2c, for example, must be caused by a recessive allele. The symbol of Individual VI-1 is shaded, which means this person has the trait. However, neither of this person's parents shows the trait. The parents must have carried the allele, because children get their alleles from their parents, but the allele isn't visible in either parent. When an allele is present but not seen in a person's phenotype, the trait must be recessive.

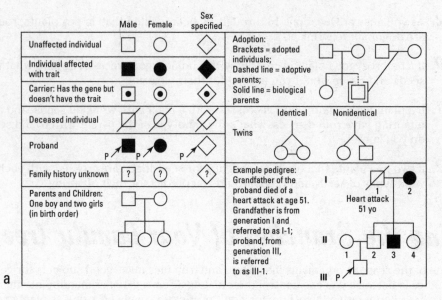

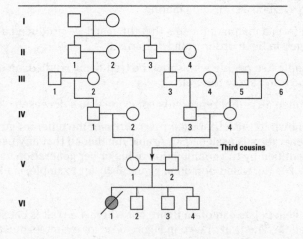

Figure 6-2:
Human
pedigrees.

Traits carried by recessive alleles often skip generations. A red-headed child may have a red-headed grandparent, for example, but blonde-haired parents. Anytime parents who *don't* show a certain trait have children who *do* show the trait, that's proof of a recessive trait.

In contrast, traits that are caused by dominant alleles tend to show up in every generation. Basically, these alleles can't hide. Every time you see an affected person, at least one of his parents should also show the trait. But people who show the dominant trait may have only one copy of the dominant allele, so they can potentially pass on recessive alleles to their children. In fact, whenever both parents show a particular trait but have kids who don't show the trait, the trait is dominant. Since the development of technology that allows scientists to manipulate and read DNA, pedigree analysis is often combined with direct analysis of a person's genetic code. For example, you may have heard that people who are at risk for certain genetic diseases because it runs in their family can be tested, in a process called *genetic screening*, to find out whether they carry the disease allele. (For more on how scientists read the genetic code, head to Chapter 7.)

Q. Does the pedigree in Figure 6-2b show a dominant or a recessive trait?

A. At first glance, the trait appears to be dominant because it shows in every generation. However, no absolute proof is immediately visible because no families have two affected parents with an unaffected child. So the only way to know for sure is to test the pedigree against each hypothesis — that the trait is dominant or that the trait is recessive.

Testing the dominant hypothesis:

1. If the trait is caused by a dominant allele, anyone who doesn't show the trait must be homozygous recessive (because the allele can't hide). So the first easy step is to write the proposed genotype aa under every open symbol.

2. Check all the shaded individuals to see whether you can assign them a genotype that doesn't conflict with the labeling you did in Step 1.

    • Individual I-1 is an unaffected male, so his genotype would be aa.

    • I-1 has unaffected children, like his son, II-4, who must also be aa. II-4 could have gotten one of his recessive alleles from his father but would need to get a second recessive allele from his mother. Is this possible?

    • II-4's mother, I-2, is affected. But if this trait is dominant, she could be heterozygous (Aa). Under this scenario, Individual II-4 could be homozygous recessive (aa). So far, the dominant hypothesis works.

If you continue testing all the individuals in this pedigree and assigning them genotypes based on the dominant hypothesis, you'll find that the hypothesis continues to work — this pedigree doesn't have any examples where it's impossible to assign a genotype that fits with the hypothesis or the observations of the family. But that doesn't prove that this particular trait is caused by a dominant allele. To see if you can be sure, you also have to test the recessive hypothesis.

Testing the recessive hypothesis:

1. To test the alternative hypothesis, you begin all over again. If the trait is caused by a recessive allele, anyone who shows the trait must be homozygous recessive. So label every person in the family that has a shaded circle or square with the genotype aa.

2. Now, go through each smaller family grouping in the tree and see whether you find any conflicts. If you begin again with Individual I-1, you see that he's unaffected. In this case, if the trait is recessive, he could be either homozygous dominant (AA) or heterozygous (Aa). So, under his symbol, write A- to show that you're not sure of his other allele.

3. I-1 has an unaffected son, II-4. II-4's mother shows the trait, which means she has to be homozygous recessive (aa). She can give only the recessive allele to II-4, so he'd have to be able to get a dominant allele from his father. This is possible, so there's no conflict yet.

If you continue in this manner through the pedigree, you'd find that no conflicts arise. So it's possible that the trait shown here is caused by a recessive allele.

The fact that the pedigree in Figure 6-2b doesn't prove either a dominant or recessive inheritance pattern for this trait illustrates a difficulty with studying human genetics: You have to work with existing pedigrees, and family sizes can be small. Because of this, geneticists who work with human traits must sometimes look at many, many pedigrees for the same trait before they can make conclusions. It's easier for recessive traits — when you see that a trait can hide, you know it's recessive. But for dominant traits, finding proof is difficult without studying many pedigrees. If a geneticist looks at many pedigrees of a particular trait and sees that it never skips generations (never hides), that's pretty suggestive that it's a dominant trait, even if none of the pedigrees contains absolute proof.

13.–17. Use the information shown in the pedigree in Figure 6-3 to answer the following questions. Use the letters *A* and *a* to represent alleles.

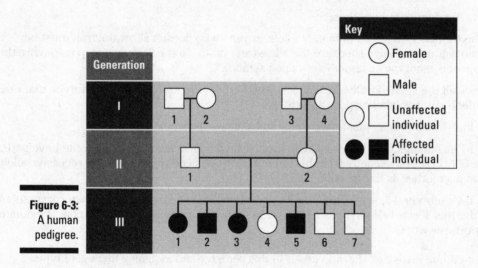

Figure 6-3:
A human
pedigree.

13. Is the trait shown in Figure 6-3 caused by a dominant or a recessive allele? How can you tell?

14. What's the gender of Individual I-2?

15. What's the phenotype of Individual III-1?

16. What's the genotype of Individual II-2?

17. What's the relationship between Individuals III-5 and III-6?

Answers to Questions on Genetics

The following are answers to the practice questions presented in this chapter.

1 The answer is **a. Chromosomes.**

2 The answer is **b. Gene.**

3 The answer is **d. Allele.**

4 The answer is **c. Locus.**

5 The answer is **b. Phenotype.**

6 The answer is **c. Heterozygous.**

7 The answer is **e. Dominant.**

8 The answer is **a. Genotype.**

9 The answer is that **the tan allele is dominant to the black allele.** You know that the pure breeding tan fly can give only messages for a tan body to the offspring and that the pure breeding black fly can give only messages for a black body. So all the F1 offspring must have one message for tan and one message for black, but they're only showing the tan allele. The tan allele hides the black allele, which means tan is dominant to black.

If the tan allele is represented by T and the black allele is represented by t, then all F1 flies have Tt for their combination of alleles. When they're mated, the F1 parents give half of their gametes the T allele and half the t allele. If you draw a Punnett square, it will look exactly like the square on the right in Figure 6-1. So the prediction for the F2 generation is the same — three dominant characteristics for every one recessive characteristic, or a 3:1 ratio of tan body to black body.

10 The answer is **half round (Rr) and half wrinkled (rr).**

The recessive parent always gives the recessive allele (r). The heterozygous parent gives the dominant allele (R) half the time and the recessive allele half the time.

11 The answer is **homozygous (RR).**

The wrinkled parent always gives the recessive allele (r). If the round parent were heterozygous, you'd expect half the offspring to be wrinkled. Because they're all round, the round parent must have only one type of allele to give — the dominant one (R).

12 The answer is **3:1 round:wrinkled.**

This is a monohybrid cross, so automatically you know that the expected ratio is 3:1 dominant to recessive. To see why this is so, check out the Punnett square to the right in Figure 6-1.

13 The answer is **a recessive allele because affected children have unaffected parents.**

The allele for the trait is hiding in the parents of generation II but reveals itself in their children in generation III.

The answer is **female because I-2 is a circle.**

 The answer is **affected (has the trait) because III-1 is shaded.**

16 The answer is **Aa.**

She is unaffected but has affected children. So she must carry the recessive allele to pass on to her kids but also have the dominant allele so that she doesn't show the trait.

17 The answer is that **III-5 and III-6 are siblings.**

Chapter 7

Taking Genetics to the DNA Level

In This Chapter

▶ Copying the message in DNA to make RNA molecules

▶ Following the instructions in mRNA to build proteins

▶ Understanding types of mutations and their consequences

Deoxyribonucleic acid (DNA) controls the structure and function of all organisms on Earth. The instructions in DNA contain the blueprints for cellular workers, like proteins, that manage the day-to-day operations of cells. How these workers function determines how cells function and thus how organisms function. When changes called *mutations* occur in the DNA of cells, the effects on the organism are sometimes, though not always, disastrous. In this chapter, you see how the human body uses the code in DNA to build RNA molecules and then protein molecules that are the main workers in cells.

Going with the Flow of Genetic Information

DNA contains the instructions for the construction of the molecules that carry out your cells' functions. These functional molecules are mostly proteins, and the instructions for creating these proteins are found in your *genes,* sections of DNA that lie along your chromosomes.

One gene equals one blueprint for a functional molecule. Because many of the functional molecules in your cells are proteins, genes often contain the instructions for building the polypeptide chains that make up proteins (for more on protein structure, head to Chapter 2). So one gene often equals one polypeptide chain.

When cells need to build a polypeptide chain, they copy the information in the genes into an RNA molecule instead of using the DNA blueprint directly (see Chapter 2 for more about RNA molecules).

The idea that information is stored in DNA, copied into RNA, and then used to build proteins is called the *central dogma of molecular biology.*

Here's an outline of the process:

> ✔ **Cells use *transcription* to copy the information in DNA into RNA molecules.** The information to build proteins is copied into a special type of RNA called *messenger RNA* (mRNA), which carries the blueprint for the protein from the nucleus to the cytoplasm, where it can be used to build the polypeptide chain (flip to Chapter 3 for a refresher on cell parts). I explain transcription in more detail in the upcoming section "Making a Copy of the Genetic Code: Transcription."

✔ **Cells use *translation* to build polypeptide chains from the information carried in mRNA molecules.** Polypeptide chains fold up into unique structures to make functional proteins. Often, more than one folded polypeptide chain comes together to make the finished protein. You discover much more about translation in the section "Decoding the Message in mRNA: Translation" later in this chapter.

Try the following questions to test your grasp of the central dogma:

1. Which cellular process uses DNA as a template and produces RNA as a product?

   a. Replication

   b. Translation

   c. Transcription

   d. Central dogma

2. To make a protein, a cell uses _____ to copy the gene and then _____ to build the protein.

   a. Replication, transcription

   b. Replication, translation

   c. Transcription, translation

   d. Translation, transcription

3. An old saying in biology is *one gene, one protein*. How would you update this saying to reflect a more complete understanding?

Making a Copy of the Genetic Code: Transcription

The order of the nitrogenous bases adenine (A), cytosine (C), guanine (G), and thymine (T) (see Chapter 2 for details on DNA structure) in a gene determines the order in which amino acids are strung together to make a protein. When your cells need to build a particular protein, the enzyme *RNA polymerase* locates the gene for that protein and makes an RNA copy of it. (RNA polymerase is shown in Step 2 of Figure 7-1 later in this section.) Because RNA and DNA are similar molecules, they can attach to each other, just like the two strands of the DNA double helix do.

When a strand of RNA pairs with a strand of DNA, the two strands arrange themselves *antiparallel* to each other, just like the two strands of the DNA double helix (see Chapter 5).

RNA polymerase opens the two strands of DNA and then slides along one strand, matching RNA nucleotides to the DNA nucleotides in the gene.

The *base-pairing rules* for matching RNA and DNA nucleotides are almost the same as those for matching DNA with DNA (see Chapter 2). The exception is that RNA contains nucleotides with uracil (U) rather than thymine (T). During transcription, RNA polymerase pairs C with G, G with C, A with T, and U with A. (Figure 7-1 shows this. Note that the new RNA strand 5'GAUC3' pairs up with the DNA sequence 3'CTAG5' in the gene.)

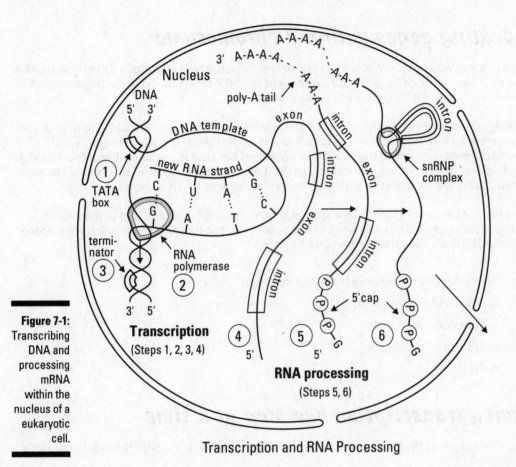

Figure 7-1:
Transcribing
DNA and
processing
mRNA
within the
nucleus of a
eukaryotic
cell.

Transcription and RNA Processing

Use the base-pairing rules to solve the following questions:

Q. If a template strand of DNA within a gene has the sequence 3'ACCGTTAGCT5', what
would be the sequence of RNA transcribed from this gene? Remember to label the RNA
strand's 5' and 3' ends.

A. Using the base-pairing rules, match the RNA nucleotides to the DNA nucleotides.
Remember that RNA molecules contain uracil (U) rather than thymine (T). Also, the two
strands are antiparallel, so the RNA strand's polarity is opposite that of the DNA strand.

DNA: 3'ACCGTTAGCT5'

RNA: 5'UGGCAAUCGA3'

4. If the template strand of DNA within a gene has the sequence 3'TTAGCATGGATCG5',
which of the following represents an RNA molecule transcribed from this gene?

a. 5'AATCGTACCTAGC3'

b. 5'CGAUCCAUGCUAA3'

c. 5'AAUCGUACCUAGC3'

d. 5'UUAGCAUGGAUGC3'

Locating genes within a chromosome

RNA polymerase locates the genes it needs to copy with the help of proteins called *transcription factors*. These proteins look for certain sequences in the DNA that mark the beginnings of genes; scientists call these sequences *promoters*.

Transcription factors find the genes for the proteins the cell needs to make and bind to the promoters so RNA polymerase can attach and copy the gene. Many promoters contain a particular sequence called the *TATA box,* so named because it contains alternating T and A nucleotides. Transcription factors bind to the TATA box first, followed by RNA polymerase. (Step 1 in Figure 7-1 depicts a gene's promoter, along with its TATA box.)

The ends of genes are marked by a special sequence called the *transcription terminator*. Transcription terminators can work in different ways, but they all stop transcription. (Step 3 of Figure 7-1 shows a transcription terminator.)

5. Where does the process of transcription begin?

    a. Origins of replication (ORI)

    b. Start codons

    c. Transcription factors

    d. Promoters

Doing transcription one step at a time

The steps of transcription are pretty straightforward (follow along with the numbered steps in Figure 7-1):

1. **RNA polymerase binds to the promoter with the help of transcription factors.**

2. **RNA polymerase separates the two strands of the DNA double helix in a small area and uses base-pairing rules to build an RNA strand that's complementary to the DNA in one strand.**

   Scientists call the DNA strand that RNA polymerase pairs against the *template strand*. RNA polymerase builds the new RNA molecule starting with its 5' end. (Scientists call the DNA strand that isn't used for base-pairing the *coding strand* because its code is the same as in the RNA molecule, but with T's in DNA and U's in RNA.)

3. **RNA polymerase reaches the termination sequence and releases the DNA.**

Q. A piece of a chromosome (DNA) that contains a gene is shown after this question. What would be the sequence of an mRNA molecule transcribed from this gene? The arrows indicate the point at which transcription begins and the direction of transcription.

            start   direction

  promoter

5'G ATTGCTATAAAACCGGATGCTACGAAT3'

3'CTAACGATATTTT GGCCTACGATGCTTA5'

A. RNA polymerase builds RNA molecules beginning with their 5' end, so it must slide along the DNA in the 3' to 5' direction. The dashed arrow shows that transcription goes to the right of the chromosome, so the bottom DNA strand must serve as the template for the RNA molecule (because it's the strand that goes from 3' to 5', in the same direction as the dashed arrow). Beginning with the start arrow, the sequence of the template DNA is

3'TTTGGCCTACGATGCTTA5'

So, using complementary base-pairing, the sequence of the RNA molecule would be

5'AAACCGGAUGCUACGAAU3'

Try putting all the steps of transcription together to solve this problem:

6. A piece of a chromosome containing a gene is shown here:

start   direction

promoter

3'GATTGCATATAAACCGGATGCTACGAAT5'

5'CTAACGTAT ATTTGGCCTACGATGCTTA3'

If this gene is transcribed to form mRNA, which of the following would be the sequence of that RNA molecule?

a. 5'UUUGGCCUACGAUGCUUA3'

b. 3'UUUGGCCUACGAUGCUUA5'

c. 5'AAACCGGAUGCUACGAAU3'

d. 3'AAACCGGAUGCUACGAAU5'

Putting on the finishing touches: RNA processing

After RNA polymerase transcribes one of your genes and produces a molecule of mRNA, the mRNA isn't quite ready to be translated into a protein. In fact, when the mRNA is hot off the presses (that is, right after it's transcribed), it's called a *pre-mRNA* or *primary transcript* because it's not quite finished.

The code within your genes isn't written in one continuous message. Instead, the code for the protein is divided into short stretches, as if the code was written in a dashed line instead of a solid line. The code for the protein is interrupted by stretches of nucleotides that don't end up in the finished protein. The pieces of code that are used to build the protein are called *exons;* the interrupting sequences are called *introns.*

Before the pre-mRNA can be translated, it has to undergo a few finishing touches via *RNA processing* (see Steps 5 and 6 in Figure 7-1):

  ✔ **The 5' cap, a protective cap, is added to the beginning of the mRNA.**

  ✔ **The poly-A tail, an extra bit of sequence, is added to the end of the mRNA.**

> ✔ **The pre-mRNA is cut to remove introns (noncoding sequences), then resealed to bring all the exons together to form the finished mRNA in a process that scientists call *splicing*.**

Now answer the following question:

7. Explain how a gene that contains 1000 nucleotides can produce a finished mRNA molecule that contains only 400 nucleotides.

8.–15. Grab some colored pencils or highlighters and use them to label the following in Figure 7-1.

8. Find and highlight the promoter with one color and label it.

9. Find the shape that represents RNA polymerase and color it in. What kind of molecule is RNA polymerase?

10. Locate the section of nucleic acid where transcription is occurring according to the base-pairing rules. Highlight the DNA one color and the RNA a different color. Label each group of four nucleotides with their 5' and 3' ends (you can find the information at the tips of the strands in the figure and then follow along until you reach the visible nucleotides).

11. Find the place where transcription will end and color it in.

12. Draw a bracket along the DNA that marks the location of the entire gene and label it with the word *gene*.

13. Find and label the primary mRNA transcript.

14. Use one color to highlight the 5' caps and another color to highlight the polyA tails.

15. Circle and label the location where splicing is occurring.

Decoding the Message in mRNA: Translation

Finished mRNA molecules leave a cell's nucleus through tiny holes in the nuclear membrane called *nuclear pores*. After the mRNA molecules reach the cytoplasm, they attach to ribosomes so they can be *translated* to produce a protein. As the strand of mRNA slides through the ribosome, the code is read three nucleotides at a time.

A group of three nucleotides in mRNA is called a *codon*. If you take the four kinds of nucleotides in RNA — A, G, C, and U — and make all the possible three-letter combinations you can, you come up with 64 possible codons.

Figure 7-2 shows the 64 codons and what they represent to a cell. Most codons represent a particular amino acid that's added to the polypeptide chain (check out Chapter 2 for details on polypeptides, proteins, and amino acids). A few special codons indicate starting and stopping points for translation.

✔ The codon AUG indicates the starting point for translation of mRNA. It also represents the amino acid *methionine*.

✔ The codons UAA, UGA, and UAG indicate stopping points for translation. They don't represent amino acids.

REMEMBER

To translate a molecule of mRNA, begin at the start codon closest to the 5' cap, divide the message into codons, and look up the codons in a table of the genetic code (like the one in Figure 7-2).

To figure out what a codon represents to a cell, uses the guidelines along the top of the table in Figure 7-2:

1. **Look first to the left side of the table and find the row marked by the first letter in the codon.**

2. **Then look to the top of the table and find the column marked by the second letter in the codon.**

3. **Look to the right side of the table and find the row marked by the third letter in the codon.**

First Letter	Second Letter				Third Letter
	U	C	A	G	
U	phenylalanine	serine	tyrosine	cysteine	U
	phenylalanine	serine	tyrosine	cysteine	C
	leucine	serine	STOP	STOP	A
	leucine	serine	STOP	tryptophan	G
C	leucine	proline	histidine	arginine	U
	leucine	proline	histidine	arginine	C
	leucine	proline	glutamine	arginine	A
	leucine	proline	glutamine	arginine	G
A	isoleucine	threonine	asparagine	serine	U
	isoleucine	threonine	asparagine	serine	C
	isoleucine	threonine	lysine	arginine	A
	methionine & START	threonine	lysine	arginine	G
G	valine	alanine	aspartate	glycine	U
	valine	alanine	aspartate	glycine	C
	valine	alanine	glutamate	glycine	A
	valine	alanine	glutamate	glycine	G

Figure 7-2:
The genetic code.

Q. What does the codon ACG represent to a cell?

A. The first letter of the codon, A, marks the third row of the table in Figure 7-2. The second letter, C, marks the second column. So the codon ACG is found in the intersection of the third row and the second column. On the right side of the table, find the letter G in the third row and then follow it over to the intersection of the third row with the second column. The word written there is *threonine,* which is the name of an amino acid.

Q. A cell translates a molecule of mRNA with the sequence 5'GCGCAUGCUGUUUCAGUGA3'. What amino acids are found in the peptide produced from this mRNA?

A. Find the start codon closest to the 5' end of the mRNA and then break up the message into codons.

5'GCGC/AUG/CUG/UUU/CAG/UGA3'

Beginning with AUG, look up the codons in Figure 7-2. The amino acids in the peptide are methionine-leucine-phenylalanine-glutamine.

UGA is a stop codon, so it doesn't represent an amino acid.

Now, you try one:

16. An mRNA molecule has the following sequence:

5'CGGCACCUUACAAUGCCAGGGUCAUAACCGAAU3'

What would be the sequence of amino acids in the peptide produced from this mRNA?

a. Arginine-histidine-leucine-threonine-methionine-proline-serine-glycine

b. Methionine-proline-glycine-serine

c. Proline-glycine-serine

d. Tyrosine-glycine-proline-serine

Deciphering mRNA codes with tRNA

When scientists, or students, want to know how to translate a particular mRNA molecule, they turn to tables like the one in Figure 7-2. Cells, of course, don't have this table. In order for your cells to decode mRNA, they need the help of an important worker: transfer RNA (tRNA).

Transfer RNA decodes the message in the mRNA. Like all RNA molecules, tRNA is made of nucleotides that can pair up with other nucleotides according to base-pairing rules.

During translation, tRNA molecules pair up with the codons in mRNA to figure out which amino acid should be added to the chain. Each tRNA has a special group of three nucleotides, called an *anticodon,* that pairs up with the codons in mRNA. Each tRNA also carries a specific amino acid. So the tRNA that has the right anticodon to pair with a specific codon adds its amino acid to the growing polypeptide chain.

Because the pairing of anticodon to codon is specific, only one tRNA can pair up with each codon. The specific relationship between tRNA anticodons and mRNA codons ensures that each codon always specifies a particular amino acid.

Use the base-pairing rules to figure out these questions about codons and anticodons.

17. What's the anticodon of the tRNA that pairs with start codons (5'AUG3')?

   a. 5'UAC3'

   b. 3'UAC5'

   c. 5'TAC3'

   d. 3'TAC5'

18. What's the anticodon of the tRNA that carries the amino acid tryptophan?

   a. 5'UGG3'

   b. 3'UGG5'

   c. 5'ACC3'

   d. 3'ACC5'

Doing translation one step at a time

Because the process of translation is fairly complicated, I break it down into three main steps for easier understanding: the beginning *(initiation),* the middle *(elongation),* and the end *(termination).* Figure 7-3 shows what happens during each of these steps:

1. **During initiation, the ribosome and the first tRNA attach to the mRNA (see #1 in Figure 7-3).**

   Ribosomes have two parts, called the *large and small subunits,* that join together around the mRNA and first tRNA.

2. **During elongation, tRNAs enter the ribosome and donate their amino acids to the growing polypeptide chain.**

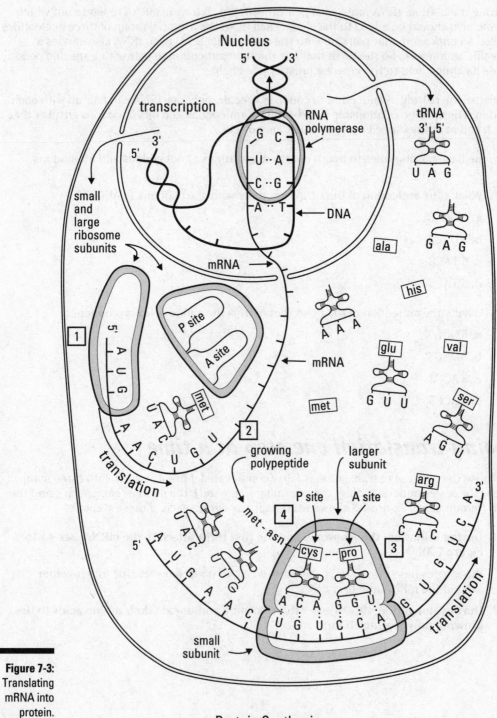

Figure 7-3:
Translating
mRNA into
protein.

Protein Synthesis

Each tRNA enters a pocket in the ribosome called the *A site* (see #3 in Figure 7-3). An adjacent pocket, called the *P site,* holds a tRNA with the growing polypeptide chain (see #4 in Figure 7-3). When a tRNA is parked in the A site and the P site, the ribosome catalyzes the formation of a *peptide bond* between the growing polypeptide chain and the new amino acid. In Figure 7-3, a bond is forming between the amino acids cysteine (cys) and proline (pro) because they're next to each other in the ribosome.

After the new amino acid is added to the growing chain, the ribosome slides down the mRNA, moving a new codon into the A site. After a new codon is in the A site, another tRNA can enter the ribosome, and the process of elongation can continue.

3. **During termination, a stop codon in the A site causes translation to end.**

The ribosome slides down the mRNA until a stop codon enters the A site. When a stop codon is in the A site, an enzyme called a *release factor* enters the ribosome and cuts the polypeptide chain free. Translation stops, and the ribosome and mRNA separate from each other.

Test your understanding of translation by tackling the following problems:

19.–27. Grab some colored pencils or highlighters and use them to label the following on Figure 7-3.

19. Find the part of the figure that represents initiation of translation. Draw a bracket next to that section and label it *initiation.*

20. In the initiation section, find the start codon, circle it, and label it.

21. Find the first tRNA. You can recognize it because its anticodon is complementary to the start codon. Label its anticodon and the amino acid it carries.

22. In the initiation section, shade in the small and large subunits of the ribosome and label which is which.

23. Find the part of the figure that represents elongation of translation. Draw a bracket next to that section and label it *elongation.*

24. In the elongation section, use one color to shade along the mRNA. Label it as mRNA.

25. In the elongation section, find the "empty" tRNAs that have already donated their amino acid to the polypeptide chain. Highlight them with a different color and label them as "empty tRNAs."

26. In the elongation section, find the tRNAs that are carrying amino acids and that will enter the A site after the ribosome slides down the mRNA. Highlight them with the same color you used in Question 25 and label them as *incoming tRNAs.*

27. Find the growing polypeptide chain and highlight it with a third color.

28.–32. Use the terms that follow to identify the structures in Figure 7-4 involved in elongation of a polypeptide during translation.

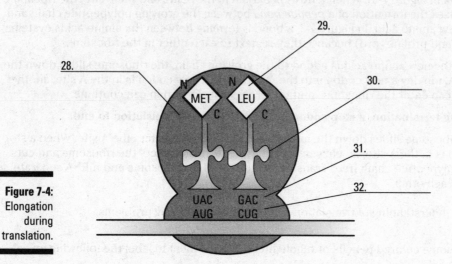

Figure 7-4:
Elongation
during
translation.

    a. tRNA

    b. Anticodon

    c. Codon

    d. Ribosome

    e. Amino acid

33.–35. Use the following DNA sequence, taken from the *coding strand* (nontemplate) of a gene, to answer the following questions.

    5'-ATG GTG ACG GGA TGA GAT-3'

33. What's the sequence of mRNA that would be transcribed from the gene?

    a. 5'-UAC CAC UGC CCU ACU CUA-3'

    b. 3'-TAC CAC TGC CCT ACT CTA-3'

    c. 5'-AUG GUG ACG GGA UGA GAU-3'

    d. 3'-UAC CAC UGC CCU ACU CUA-5'

34. In order from left to right, what are the anticodons that would pair with the mRNA transcribed from this DNA?

    a. 5'-UAC CAC UGC CCU ACU CUA-3'

    b. 3'-TAC CAC TGC CCT ACT CTA-3'

    c. 5'-AUG GUG ACG GGA UGA GAU-3'

    d. 3'-UAC CAC UGC CCU ACU CUA-5'

35. In order from left to right, what are the amino acids in the peptide that would be produced by transcribing and translating this gene?

  a. Methionine-valine-threonine-glycine

  b. Tyrosine-histidine-cysteine-proline-threonine-leucine

  c. Valine-threonine-glycine

  d. Isoleucine-serine-serine-arginine-histidine-histidine

Measuring the Impact of Mutations

If a mistake in a strand of DNA goes undetected or unrepaired, the mistake becomes a mutation. A *mutation* is a change from the original DNA strand; in other words, the nucleotides aren't in the order that they should be.

Mutations in DNA lead to changes in RNA, which can lead to changes in proteins. When proteins change, the function of cells and the traits of organisms can also change.

Mutations usually happen as the DNA is being copied during DNA replication (see Chapter 5 for a description of DNA replication). Two main types of mutations occur:

✔ **Spontaneous mutations:** These result from uncorrected mistakes by *DNA polymerase,* the enzyme that copies DNA. DNA polymerase is a very accurate enzyme, but it's not perfect. In general, DNA polymerase makes one mistake for every billion base pairs of DNA it copies.

✔ **Induced mutations:** These result from environmental agents such as radiation or certain chemicals that increase the rate of change of DNA. Anything that increases the rate of DNA changes a *mutagen.*

Mutations happen randomly, so their outcomes are random. In other words, an organism can't make a particular mutation happen in order to solve a problem it faces. Most mutations that have an effect on an organism result in something not working correctly, and therefore the mutations have a negative impact. But occasionally, a mutation happens that gives an organism an advantage in its particular environment. When this happens, the organism may have more offspring than its competition, and the mutation may spread in the population. (For more on this, head to Chapter 11 and check out the subject of evolution.)

When mutations occur during DNA replication, some daughter cells formed by mitosis or meiosis inherit the genetic change. The types of mutations these cells inherit can be divided into three major categories:

✔ **Base substitutions:** When the wrong nucleotides are paired together in the parent DNA, a *base substitution* occurs. If the parent DNA molecule has a nucleotide containing thymine (T), DNA polymerase should bring in a nucleotide containing adenine (A) for the new strand of DNA. However, if DNA polymerase brings in a nucleotide with guanine (G) by mistake, that's a base substitution. Because just one nucleotide is changed, the mutation is called a *point mutation.* The effect of point mutations ranges from nothing to severe:

- **Silent mutations** have no effect on the protein or organism. The genetic code is *redundant* so that multiple codons may represent one amino acid. Because of this, changes in DNA may lead to changes in mRNA that don't cause changes in the protein. (For example, take another look at Figure 7-2 and see how many codons represent the amino acid leucine.)

- **Missense mutations** change the amino acids in the protein. Changes in DNA can change the codons in mRNA, leading to the substitution of different amino acids into a polypeptide chain.

- **Nonsense mutations** introduce a stop codon into the mRNA, preventing the protein from being made. If the DNA changes so that a codon in the mRNA becomes a stop codon, the polypeptide chain gets cut short.

✔ **Deletions:** When DNA polymerase fails to copy the entire DNA in the parent strand, that's a *deletion.* If nucleotides in the parent DNA are read but the complementary bases aren't inserted, the new strand of DNA is missing nucleotides. If one or two nucleotides are deleted, the codons in the mRNA are skewed from their proper threes, and the resulting polypeptide chain is altered. Mutations that change the way in which the codons are read are called *frameshift mutations.*

✔ **Insertions:** When DNA polymerase slips and copies nucleotides in the parent DNA more than once, an *insertion* occurs. Just like deletions, insertions of one or two nucleotides can cause frameshift mutations that greatly alter the polypeptide chain.

Q. A coding (nontemplate) strand of DNA has the following sequence:

5'-CGG-TTA-GCC-ACG-GAC-TAA-3'

If a mutation occurs in the DNA, changing the fourth codon from ACG to TCG, what type of mutation is this? What effect will it have on the polypeptide?

A. The coding strand of DNA contains the same code as the mRNA that's transcribed from the template strand. So to figure out the normal mRNA, just replace all the T's with U's. This gives you:

Normal mRNA 5'-CGG-UUA-GCC-ACG-GAC-UAA-3'

The fourth codon in the mRNA is ACG, which represents the amino acid threonine.

If the fourth codon in the DNA mutates to TCG, the fourth codon in the mRNA will change to UCG. This mutated codon represents the amino acid serine. The amino acid changed as a result of the mutation, so this is a missense mutation.

Your turn! Try some problems and see if you can figure out the consequences of these mutations.

36. The coding strand of DNA from a gene has the following sequence:

5'-ATG-TCG-ACG-TCG-ACG-3'

If the seventh nucleotide is lost from this strand because of a deletion, what type of mutation will result?

a. Missense affecting one amino acid only

b. Frameshift resulting in extensive missense

c. Silent mutation

d. Nonsense mutation

37. The coding strand of DNA from a gene has the following sequence:

5'-ATG-TCG-ACG-TCG-ACG-3'

If the sixth nucleotide in this strand changes from a G to a T, what type of mutation will result?

a. Missense affecting one amino acid only

b. Frameshift resulting in extensive missense

c. Silent mutation

d. Nonsense mutation

38. The coding strand of DNA from a gene has the following sequence:

5'-ATG-TCG-ACG-TCG-ACG-3'

If the eleventh nucleotide in this strand changes from a C to an A, what type of mutation will result?

a. Missense affecting one amino acid only

b. Frameshift resulting in extensive missense

c. Silent mutation

d. Nonsense mutation

Answers to Questions on the Genetic Code

The following are answers to the practice questions presented in this chapter.

1 The answer is **c. Transcription.**

Transcription uses DNA as a template and produces RNA as a product. In comparison, replication uses DNA as a template and produces DNA as a product, and translation uses mRNA as a template and produces polypeptides as a product.

2 The answer is **c. Transcription, translation.**

3 Some proteins assemble from more than one folded polypeptide chain in order to create the final functional molecule. So a more accurate statement would be *one gene, one polypeptide.* But even this statement isn't 100 percent accurate because some genes contain codes for functional molecules other than proteins, such as tRNA, rRNA, and other types of RNA molecules. So the most precise statement is probably *one gene, one gene product.*

4 The answer is **c. 5'AAUCGUACCUAGC3'.**

RNA polymerase reads the template strand of DNA from its 3' end to its 5' end, using the base-pairing rules to build an antiparallel mRNA molecule.

5 The answer is **d. Promoters.**

Compare this to replication, which begins at origins of replication, and translation, which begins at start codons.

6 The answer is **a. 5'UUUGGCCUACGAUGCUUA3'.**

The start arrow shows where transcription should begin, and the dashed arrow shows the direction of transcription. RNA polymerase reads the template strand from its 3' end to its 5' end, so in this case it would have to read the top strand (because it's the one that goes from 3' to 5' in the direction of travel). You can look at the top strand and use base-pairing rules to figure out the complementary mRNA molecule or you can use the fact that the coding (nontemplate) strand contains the same code as the mRNA.

7 Genes contain introns that interrupt the coding regions (exons) in the gene. A gene is transcribed into pre-mRNA and then introns are edited out to make a finished mRNA, which shortens the final transcript. In this case, 600 nucleotides were edited out of the pre-mRNA.

8 The promoter can be recognized by the TATA box that's marked with #1.

9 The RNA polymerase is shaped like a guitar pick in this picture and is marked near #2. RNA polymerase is an enzyme (you can tell because its name ends in –ase) and a protein.

10 The DNA sequence is 5'GATC3'. The RNA sequence is 3'CUAG5'.

11 Transcription ends at the terminator sequence, which is marked with #3 in the figure.

12 The gene is located from the promoter (marked at #1) to the terminator (marked at #3).

13 The primary mRNA transcript is marked with #4.

14 The 5' caps have 3 phosphates — each represented as a P in a circle — and a guanine (G). You can see them in both strands marked by #5 and #6. The poly-A tails are the long strings of repeated A's visible in both strands marked by #5 and #6.

15 Splicing is occurring in the strand marked #6 at the location labeled *snRNP complex*.

16 The answer is **b. Methionine-proline-glycine-serine.**

mRNA is translated beginning at its 5' end and moving toward its 3' end. So begin at the 5' end and look for the first start codon (AUG). After you find the start codon, divide all the nucleotides into groups of three. Beginning with the start codon, look up each codon in the codon dictionary shown in Figure 7-2. UAA is a stop codon and marks the end of translation.

17 The answer is **b. 3'UAC5'.**

The anticodons of tRNA molecules bind to the codons in mRNA following base-pairing. The two types of RNA molecules attach to each other in an antiparallel orientation using hydrogen bonds, just like the two strands in a DNA double helix. The start codon is 5'AUG3', so the anticodon that can attach to this codon is 3'UAC5'.

18 The answer is **d. 3'ACC5'.**

If you look in the codon dictionary in Figure 7-2, you see that tryptophan is only represented by one codon: 5'UGG3' (codons are read 5' to 3'). The anticodon to this codon is 3'ACC5'.

19 Initiation is occurring at the section of the figure marked with #1. You can see the two separate pieces of the ribosome and the first tRNA.

20 The start codon is the AUG that's tucked inside of the separate small subunit.

21 The first tRNA is in between the separate pieces of the ribosome. Its anticodon is 3'UAC5' (you can figure out the 3' and 5' ends because the anticodon pairs with the start codon, so it must be opposite in polarity). The abbreviation *met* inside the rectangle represents the amino acid methionine.

22 The small subunit binds to the mRNA first during initiation, so it's the piece that's already attached to the mRNA (and the start codon). The large subunit shows part of the P and A sites inside it.

23 Elongation is occurring along the bottom of the figure where the completed ribosome is attached around the mRNA.

24 The mRNA is the long string of nucleotides that begins 5'AUGAAC . . . 3'.

25 Two empty tRNAs are just above the 5' end of the mRNA. One has the anticodon 3'UAC5'; the other has the anticodon 3'UUG5'. (How did I know which ends were 3' and which were 5'? See the answer to Question 21 if you're not sure.)

26 Incoming tRNAs are on the right side of the ribosome, near #3. One has the anticodon 3'CAC5' and is carrying the amino acid arg (arginine); the other has the anticodon 3'AGU5' and is carrying the amino acid ser (serine).

27 The growing polypeptide chain is labeled at #2. You can recognize it because it's a string of abbreviations for amino acids (met-asn-cys-pro).

28—32 The following is how Figure 7-4 should be labeled:

28. **d. Ribosome;** 29. **e. Amino acid;** 30. **a. tRNA;** 31. **b. Anticodon;** 32. **c. Codon.**

33 The answer is **c. 5'-AUG-GUG-ACG-GGA-UGA-GAU-3'.**

The coding strand contains the same sequence as the mRNA that would be transcribed from this gene, so to figure out the mRNA, all you have to do is replace the T's in the codon strand with U's.

34 The answer is **d. 3'-UAC-CAC-UGC-CCU-ACU-CUA-5'.**

After you figure out the mRNA in Question 16, you can use base-pairing rules to figure out the tRNA.

35 The answer is **a. Methionine-valine-threonine-glycine.**

Go back to the mRNA you figured out for Question 16 and then use Figure 7-2 to look up the codons (remember that the codon dictionary in Figure 7-2 shows codons, not anticodons, so you always look up the mRNA message, *not* the tRNAs). To translate mRNA, you always start with the start codon closest to the 5' end, but in the mRNA from Question 16, the first codon is a start codon, so look up codons right from the beginning of the chain.

36 The answer is **b. Frameshift resulting in extensive missense.**

The mRNA transcribed from this gene would be a mirror image of the coding strand, with the T's swapped out for U's. So the normal mRNA would be 5'-AUG-UCG-ACG-UCG-ACG-3'. If you look up the normal mRNA in the codon dictionary (Figure 7-2), you see that the normal peptide would read methionine-serine-threonine-serine-threonine.

Now, if the seventh nucleotide is lost, the mRNA changes to 5'-AUG-UCG-CGU-CGA-CG-3'. If you look up this mRNA, you find that everything after the second codon is changed. (Remember, frameshift mutations are mutations in which the codons get shifted because of insertions or deletions.)

37 The answer is **c. Silent mutation.**

This coding strand is the same as the one in Question 18, so you can refer back to that answer for the normal mRNA and peptide sequence. If the sixth nucleotide in the DNA changes from a G to a T, the mRNA changes to 5'-AUG-UCU-ACG-UCG-ACG-3'. The original second codon, UCG, represents serine, and so does the changed codon.

38 The answer is **d. Nonsense mutation.**

This coding strand is the same as the one in Question 18, so you can refer back to that answer for the normal mRNA and peptide sequence. If the eleventh nucleotide in the DNA changes from a C to an A, the mRNA changes to 5'-AUG-UCG-ACG-UAG-ACG-3'. This changes the fourth codon to a stop codon.

Chapter 8

Going Straight to the Source with DNA Technology

*B*iologists today can go right to the source of information about living things, reading and manipulating the genetic code in DNA to solve problems in science, medicine, and society. An entire field, called *molecular biology,* has developed as a result of new tools and techniques that allow scientists to work directly with the smallest components of living things. In this chapter, I walk you through a few of the most fundamental — and most important — tools that molecular biologists use when working with DNA. I also give you a chance to put yourself in the shoes of the scientists and practice solving a few problems using these techniques.

Discovering the Power of DNA Technology

Over the past 30 years, a revolution has been taking place in the sciences of biology and medicine. At the heart of this revolution is the ability to go directly to the source — DNA — and read the genetic code of life itself. You've seen evidence of this revolution if you watch detective shows or read books about crime that feature forensic science and DNA evidence. The revolution in biology has also affected your life if you've ever wondered about genetically modified organisms (GMOs) in your food supply, if you or someone you know takes insulin as treatment for diabetes, or if you've ever heard of the human genome project. All these components of modern life began with scientists using *DNA technology,* the tools and techniques used for reading and manipulating the DNA code.

Scientists use DNA technology to try and solve human problems. Scientists can introduce genes from one organism into another organism, causing the second organism to make new proteins (for the connection between DNA and proteins, check out Chapter 7). When scientists combine DNA from two sources, they say that the new organism is *recombinant.* When scientists alter the genetic code of an organism, like they do with GMOs, they call the process *genetic engineering.* Scientists use genetic engineering for many reasons, including the following:

✔ To introduce genes for pest resistance or increased nutrition into crop plants.

✔ To introduce human genes into bacteria so that bacteria can make human proteins like insulin for medicinal use.

✔ To introduce normal genes into the cells of people with genetic diseases to help them function normally. (This medical procedure is called *gene therapy.*)

In addition to genetic engineering, scientists use DNA technology to create *DNA fingerprints,* which allow scientists to compare the DNA of one organism with another. DNA fingerprinting is useful in many situations, including:

✔ Helping to identify relatives

✔ Comparing DNA left at crime scenes with that of suspects

✔ Creating a reliable marker, or genetic pattern, of a specially bred animal or plant, such as a famous racehorse or a genetically engineered crop plant, in order to track its descendants

The DNA code is also a source of information about genes and how they control the traits of organisms. By reading DNA, scientists hope to further their understanding of all life on Earth and of human diseases:

✔ As part of the *human genome project,* a global team of scientists read the entire DNA code, called the *genome,* from human cells. The project, which was completed in 2003, basically created a map showing the location of all the genes on human chromosomes, which is enormously helpful to scientists and doctors who want to understand how human genes control body function.

✔ People who are at risk for inheriting genetic diseases may seek *genetic screening* to determine whether they carry genes that put them at greater risk.

Try this question and then read on to find out more about the techniques scientists use to manipulate and decode DNA.

1. What do scientists call the process of giving people with a genetic disease a normal copy of a gene to improve their health?

a. DNA fingerprinting

b. Genetic screening

c. Gene therapy

d. Reading a genome

Cutting DNA with Restriction Enzymes

Scientists use *restriction enzymes* to cut DNA into smaller pieces so they can analyze and manipulate DNA more easily. Each restriction enzyme recognizes and can attach to a certain sequence on DNA called a *restriction site.* You can think of restriction enzymes as little molecular scissors that slide along the DNA and cut the sugar-phosphate backbone wherever they find their restriction site (see Chapter 2 for more on the structure of DNA).

Figure 8-1 shows how a restriction enzyme can make a cut in a circular piece of DNA and turn it into a linear piece.

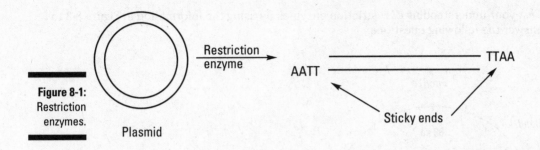

Figure 8-1:
Restriction
enzymes.

Some restriction enzymes make a straight cut through the DNA backbone, while others, like the one shown in Figure 8-1, make staggered cuts. The enzymes that make staggered cuts leave small pieces of single-stranded DNA at the ends of the fragments they cut. Scientists call these single-stranded pieces *sticky ends* because they have complementary sequences to each other and tend to stick together by hydrogen bonds.

You can get two different pieces of DNA to stick together if you cut them both with a restriction enzyme that makes sticky ends. The two pieces tend to attach to each other, making it possible to combine them into a *recombinant DNA* molecule that has DNA from two sources.

Q. A small, linear piece of viral DNA is shown in Figure 8-2. The viral DNA contains restriction sites for two different restriction enzymes, called EcoR1 and HindIII. The locations of the restriction sites are marked with arrows on the picture of the viral DNA. The length of DNA is given in kb, which stands for *kilobase pairs*. (*Kilo* means *thousand,* so one kilobase is 1000 base pairs of DNA.) If you used just EcoR1 to cut a sample that contained many copies of this piece of DNA, how many differently sized pieces of DNA would result? What size would they be? If you cut the DNA with just HindIII, what would result? And what would happen if you cut the DNA with both enzymes?

Figure 8-2:
Example
of cutting
DNA with
restriction
enzymes.

| EcoR1 | HindIII |
| 0 | 15 kb | 22.5 kb | 30 kb |

A. If you cut the DNA with just EcoR1, the DNA would be cut right in the middle. All the pieces would be the same size, which would be 15 kb long.

If you cut the DNA with just HindIII, the DNA would be cut at the 22.5 kb mark. Half of the DNA pieces would be 22.5 kb long, and the other half would be 7.5 kb long (30 kb – 22.5 kb = 7.5 kb).

If you cut the DNA with both restriction enzymes, you'd get two cuts — one at the halfway point of 15 kb and the other at the 22.5 kb mark. So, half of the pieces would be 15 kb long. The other half would all be 7.5 kb long because the cut at 22.5 kb would be right in the middle of the second half of the DNA.

Test your understanding of restriction enzymes by using the information in Figure 8-3 to answer the following questions.

Figure 8-3: Practicing with restriction enzymes. A circular piece of DNA, called a *plasmid,* is 80 kb long. The marks on the plasmid indicate restriction sites for the enzymes EcoR1 and BamH1.

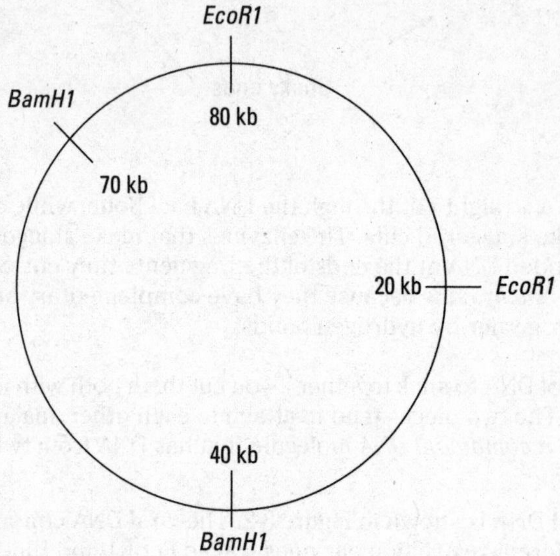

2. What types of DNA fragments would result if you were to cut a DNA sample containing many copies of the plasmid shown in Figure 8-3 with the restriction enzyme EcoR1?

   a. ½ 20 kb, ¼ 30 kb, ¼ 10 kb

   b. ½ 20 kb, ½ 60 kb

   c. ½ 30 kb, ½ 50 kb

   d. All 80 kb

3. What types of DNA fragments would result if you were to cut the plasmid shown in Figure 8-3 with both restriction enzymes, EcoR1 and BamH1?

   a. ½ 20 kb, ¼ 30 kb, ¼ 10 kb

   b. ½ 20 kb, ½ 60 kb

   c. ½ 30 kb, ½ 50 kb

   d. All 80 kb

4. What types of DNA fragments would result if you were to cut the plasmid shown in Figure 8-3 with the restriction enzyme HindIII?

   a. ½ 20 kb, ¼ 30 kb, ¼ 10 kb

   b. ½ 20 kb, ½ 60 kb

   c. ½ 30 kb, ½ 50 kb

   d. All 80 kb

Separating Molecules with Gel Electrophoresis

Scientists use *gel electrophoresis* to separate molecules based on their size and electrical charge. Gel electrophoresis can separate fragments of DNA that were cut with restriction enzymes, creating a visual map of fragment size that's easy to interpret. Or scientists may use gel electrophoresis to separate a protein they want to study from other proteins in a cell. One of the advantages of gel electrophoresis is that scientists can separate several samples side by side so they can compare them. This comparison of separated DNA molecules is the basic method behind the *DNA fingerprints* that forensic scientists use to compare samples from crime scenes with those of suspects.

Scientists conduct gel electrophoresis by inserting molecules such as DNA into little pockets called *wells* within a slab of gel (see Figure 8-4). They then place the gel in a box called an *electrophoresis chamber* that's filled with a salty, electricity-conducting buffer solution.

The DNA molecules, which have a negative charge, move toward the gel box's positive electrode because opposite charges attract. When the scientists run an electrical current through the gel, the gel becomes like a racetrack for the DNA molecules as they try to get to the positively charged end of the box.

When the power is turned off, all the DNA molecules stop where they are in the gel, and the scientists stain them. The stain sticks to the DNA, creating stripes called *bands*. Each band represents a collection of DNA molecules that are the same size and stopped in the same place in the gel.

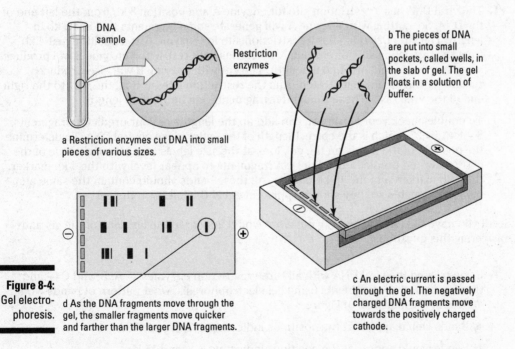

DNA sample

Restriction enzymes

b The pieces of DNA are put into small pockets, called wells, in the slab of gel. The gel floats in a solution of buffer.

a Restriction enzymes cut DNA into small pieces of various sizes.

Figure 8-4: Gel electrophoresis.

d As the DNA fragments move through the gel, the smaller fragments move quicker and farther than the larger DNA fragments.

c An electric current is passed through the gel. The negatively charged DNA fragments move towards the positively charged cathode.

In order to work with the information from the gel more easily, scientists can make an identical copy of the gel by transferring the DNA molecules to a thin sheet of nylon or nitrocellulose, a strong but flexible material that binds to DNA. This procedure is called making a *blot* of the gel. A blot on a thin, flexible material can be handled, whereas the original slab of gel can crack and break.

Q. Figure 8-5 shows a linear piece of viral DNA that's 27 kb long and has restriction sites for the enzymes A, B, and C. After the viral DNA was cut with various combinations of restriction enzymes, the resulting DNA fragments were separated using gel electrophoresis. Based on the information given in the figure, what pattern of bands would you predict in the lane of the gel marked A + B, which would be loaded with many copies of viral DNA cut with both enzymes A and B?

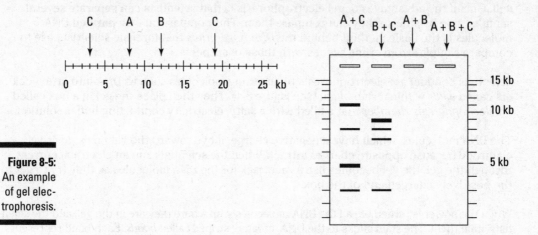

Figure 8-5:
An example
of gel elec-
trophoresis.

A. The viral DNA has a restriction site for enzyme A at a position 8 kb from the left end of the DNA. So, cutting with enzyme A will generate some fragments that are 8 kb in length. The viral DNA also has a restriction site for enzyme B at a position just 4 kb from the restriction site for enzyme A, so cutting with restriction enzyme B will produce some fragments that are 4 kb in length. The cut with enzyme B will also produce remainder fragments that stretch from the restriction site for B all the way to the right end of the viral DNA. These remainder fragments will be 15 kb in length.

To double-check your work, you can add up the lengths of your predicted fragments: 8 + 4 + 15 = 27, which is the correct length of the entire piece of viral DNA. To determine the position of the bands on the gel, look at the size labels along the right side of the gel. You would predict a band of DNA fragments to appear level with the 4 kb marker, the 8 kb marker, and the 15 kb marker. All these bands should contain the same numbers of fragments so they should appear in equal thickness on the gel.

See if you have the idea by figuring out what would appear in the last lane of the gel and answering this question.

5. If you treated the viral DNA with all three restriction enzymes — A, B, and C — and then separated the fragments using gel electrophoresis, what pattern of bands would appear in the final lane of Figure 8-5?

a. Bands would appear at the positions indicating 3, 5, 8, and 11 kb

b. Bands would appear at the positions indicating 4, 8, and 15 kb

c. Bands would appear at the positions indicating 3, 7, 8, and 9 kb

d. Bands would appear at the positions indicating 3, 4, 5, 7, and 8 kb

Use the information in Figure 8-5 to answer the following questions.

6. The open rectangles at the top of the gel in Figure 8-5 represent the wells. Towards which end of the gel would the positive electrode be located?

   a. The top

   b. The bottom

7. The open rectangles at the top of the gel in Figure 8-5 represent the wells. If you were to run a sample of uncut viral DNA through this gel for the same amount of time that the other samples were run, at which end of the gel would you expect to find your band?

   a. The top, near the wells

   b. The bottom, away from the wells

Copying DNA with PCR

The *polymerase chain reaction* (PCR) is a process that can turn a single copy of a gene into more than a billion copies in just a few hours. It gives medical researchers the ability to make many copies of a gene whenever they want to genetically engineer something or give forensic scientists enough DNA to work with when they're dealing with samples from a crime scene or mass grave.

PCR targets the gene to be copied with *primers,* single-stranded DNA sequences that are complementary to sequences next to the gene to be copied.

To begin PCR, the DNA sample that contains the gene to be copied is combined with thousands of copies of primers that frame the gene on both sides (see Figure 8-6). DNA polymerase uses the primers to begin DNA replication and copy the gene (refer to Chapter 5 for more on DNA replication).

A special heat-resistant DNA polymerase, called *Taq polymerase,* is used for PCR so that scientists can use cycles of heating and cooling to separate the strands of DNA for replication without damaging the polymerase.

The basic steps of PCR are repeated over and over until you have billions of copies of the DNA sequence between the two primers.

PCR works a little like chain e-mails. If you get a chain e-mail and send it on to two friends, who each send it on to two of their friends, and so on, pretty soon everyone has seen the same e-mail. In PCR, first a DNA molecule is copied, then the copies are copied, and so on, until you have 30 billion copies in just a few hours.

Here's a question to test your understanding of the PCR reaction.

8. You work in a forensic crime lab. The investigators from a crime scene bring you a few hairs with attached skin cells that they collected at the scene. They ask you to compare the DNA from the skin cells with the DNA of their two main suspects. You only have a few cells from the crime scene, so you know you'll need to do PCR to amplify the DNA sample. You extract the DNA from the skin cells and get ready to run the PCR reaction. Into a test tube containing a buffer solution, you put the DNA, some nucleotides, and the enzyme *Taq polymerase*. You're just about to put the tube in your PCR machine when you realize you've forgotten something. What did you forget to put in your PCR tube? What would happen if you ran your reaction without adding anything else?

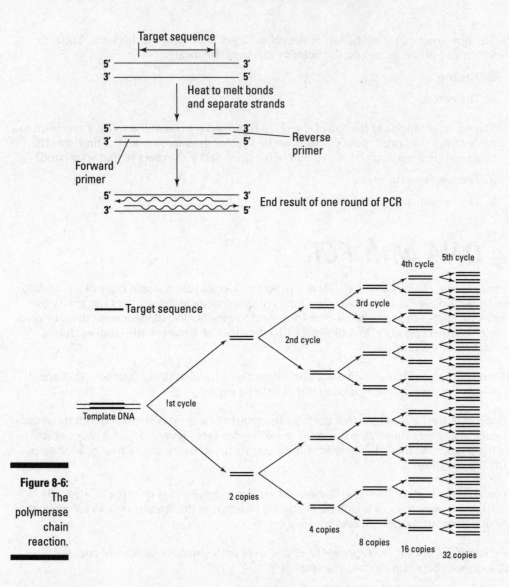

Figure 8-6:
The polymerase chain reaction.

Reading a Gene with DNA Sequencing

DNA sequencing, which determines the order of nucleotides in a DNA strand, allows scientists to read the genetic code so they can study the normal versions of genes. It also allows them to make comparisons between normal versions of a gene and disease-causing versions of a gene. After they know the order of nucleotides in both versions of a gene, they can identify which changes in the gene cause the disease.

DNA sequencing relies upon a special kind of nucleotide, called ddNTP (short for *dideoxyribonucleotide triphosphate*). ddNTPs are somewhat similar to regular DNA nucleotides, but they're different enough that they stop DNA replication (see Chapter 5 for details on DNA replication). When a ddNTP is added to a growing chain of DNA, DNA polymerase can't add any more nucleotides. DNA sequencing uses this chain interruption to determine the order of nucleotides in a strand of DNA.

Most DNA sequencing done today is *cycle sequencing* (shown in Figure 8-7), a process that works like PCR to create many copies of the gene to be sequenced. But, unlike PCR, ddNTPs as well as regular nucleotides are added to the mix. So, as the DNA polymerase works away making many copies of the gene, every now and then it grabs a ddNTP (shown as white rectangles in Figure 8-7) instead of a regular nucleotide (shown as black rectangles in Figure 8-7). After a ddNTP is added to a growing DNA molecule, replication is stopped. The end result of cycle sequencing is lots of partial copies of the gene to be sequenced, all of which are stopped at different points and are therefore different lengths.

After the partial copies are made, scientists load them into a machine that uses gel electrophoresis to put the copies into order by size (see Figure 8-7). As the partial sequences pass through the machine, a laser reads a fluorescent tag on each ddNTP, which reveals the order of nucleotides in the gene.

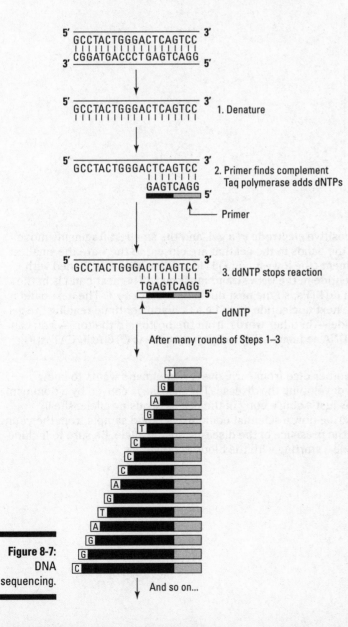

Figure 8-7:
DNA
sequencing.

Test your ddNTP reading comprehension with this practice question:

Q. A scientist wants to determine the sequence of a gene associated with a genetic disease. She takes samples of DNA from people who have the disease and puts them into four separate test tubes, each containing nucleotides, DNA polymerase, and primers. Then, into each tube, she adds one type of dideoxynucleotide — ddATP, ddGTP, ddCTP, or ddTTP — so that each tube will produce DNA fragments that end with a particular nucleotide (A, G, C, or T). After running the sequencing reactions, she loads the results of each tube into a gel and separates the fragments using gel electrophoresis. Based on the results in her gel, shown in Figure 8-8, what is the sequence of the gene?

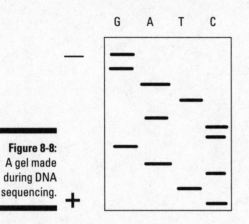

Figure 8-8:
A gel made during DNA sequencing.

A. DNA travels toward the positive electrode in a gel, and the smallest fragments move the greatest distance, so the bands in the gel that are closest to the + are the smallest bands. The shortest fragment (the lowest band) is in the lane that was loaded with ddCTPs, so the first nucleotide in the DNA strand must be C. The next band is in the lane that was loaded with ddTTPs, so the next nucleotide must be T. The next band is again in the C lane, so the next nucleotide must be a C. If you continue reading the gel from the + side to the – side — in other words, from the bottom to the top — you can figure out the rest of the DNA sequence. The entire sequence is CTC AGC CAT AGG.

9. A young woman whose mother died from early onset Alzheimer's wants to know whether she is at risk for developing the disease. The disease is caused by a dominant allele, so if the woman has just a single copy of the disease-causing allele, she'll develop the disease. Describe how a scientist could use a blood sample from the young woman to screen her for the presence of the disease-causing allele. Be sure to include all the necessary techniques, starting with the blood sample.

Answers to Questions on DNA Technology

The following are answers to the practice questions presented in this chapter.

 The answer is **c. Gene therapy.**

 The answer is **b. ½ 20 kb, ½ 60 kb.**

*Eco*R1 would make two cuts in each plasmid.

3 The answer is **a. ½ 20 kb, ¼ 30 kb, ¼ 10 kb.**

If you cut the plasmids with both restriction enzymes, you'd end up with four fragments per plasmid — two 20kb in length, one 30kb, and one 10kb.

4 The answer is **d. All 80 kb.**

If you cut the plasmid with the restriction enzyme *Hind*III, you wouldn't get any cuts because there aren't any restriction sites for this enzyme (restriction enzymes are specific to the sites they recognize).

5 The answer is **d. Bands would appear at the positions indicating 3, 4, 5, 7, and 8 kb.**

If you cut the viral DNA with all three enzymes, you'll make four cuts, resulting in five fragments.

 The answer is **b. The bottom.**

DNA, which is negatively charged, is loaded into the wells. The positive electrode is located at the far end, away from the DNA, so that it will attract the DNA from a distance, pulling it through the gel.

7 The answer is **a. The top, near the wells.**

Uncut pieces of viral DNA would be longer than any of the fragments made by restriction enzymes (they'd be the entire 27 kb long), so they would travel less far than any of the fragments.

8 The answer is **the primers.**

The primers bracket the region of DNA you want to copy and are needed to get replication started. Without them, no copies would be made.

9 The scientist would extract DNA from the sample of blood and then use either DNA sequencing or restriction enzymes to determine whether the young woman carries the disease-causing allele.

DNA sequencing method: The scientist reads the sequences of the young woman's alleles and compares them to the known sequences of the normal and disease alleles. The scientist separates the young woman's DNA sample and places it into four tubes. Each tube contains primers for the Alzheimer's gene and lots of nucleotides. One of each tube contains a different fluorescently labeled ddNTP: one with ddATP, one with ddCTP, one with ddGTP, and one with ddTTP. The scientist places the tubes into a cycle sequencer, where they go through many rounds of DNA replication. Then the scientist runs the samples through a gel that separates them by size. The computer detects the fluorescent signal at the end of each piece of DNA from smallest to largest and uses it to reconstruct the DNA sequences for the young woman's alleles.

Restriction enzyme method: The scientist uses PCR on the DNA from the blood to make many copies of the gene associated with Alzheimer's. Then the scientist uses a restriction enzyme that's known to cut differently within the normal and disease forms of the gene. The scientist cuts the young woman's DNA sample with the restriction enzyme and then separates the DNA using gel electrophoresis. The scientist compares the sizes of the DNA fragments produced by the young woman's DNA to the known pattern generated by the normal and disease alleles.

Part III
Making Connections with Ecology and Evolution

In this part . . .

In every environment on Earth, living things interact with one another to find the things they need to survive — food, water, homes, and mates. On the most fundamental level, organisms interact with one another to obtain the matter and energy they need to grow and reproduce. All life on Earth is related and connected in one big family tree.

Organisms that are successful in getting what they need to survive can reproduce, passing on the traits that made them successful to their offspring. Over time, shifts in the Earth's environment change the requirements for success, causing shifts in populations of living things.

In this part, I explore the diversity of life on Earth and explain the connections among living things in space and time through the fields of ecology and evolution.

Chapter 9

Organizing the Living World

Life on Earth is incredibly diverse and abundant. Biologists look at all this diversity and ask questions about how it came to be. They also look to the fossil record and wonder about organisms that were alive in the past. One of the big questions that biologists ask concerns the relationships among organisms alive today and organisms from the past. Scientists look at the characteristics of organisms and try to create family trees, called *phylogenetic trees,* that reflect these relationships. In this chapter, I introduce you to the methods scientists use to organize and classify living things and give you some practice interpreting phylogenetic trees.

Examining Relationships

Much like you'd draw a family tree to show the relationships among your parents, grandparents, and other family members, biologists draw *phylogenetic trees* to show the relationships of organisms within a group.

The more characteristics two organisms have in common with each other, the more closely related they are. Characteristics that organisms have in common are called *shared characteristics.*

Although you probably know how your family members are related to one another, biologists have to use clues to figure out the relationships among living things. The types of clues they use to figure out relationships include

▶ **Physical structures:** The structures that biologists use for comparison may be large, like feathers, or very small, like a cell wall (flip to Chapter 3 for more on cell walls). Biologists can examine physical structures on living organisms and in the fossil record. Structures may be similar for two reasons:

  • *Homologous structures* are similar because they evolved from structures in a common ancestor. A human hand and a bat wing are homologous structures because they have all the same bones and clearly evolved from the same ancestral structure.

- *Analogous structures* are similar because they evolved to serve a similar function, but they don't reflect an evolutionary relationship. A fish fin and a dolphin fin are analogous structures. Both evolved for swimming through water, but they didn't evolve from a common ancestral structure.

  Analogous structures aren't useful for constructing phylogenetic trees because they don't reflect evolutionary relationships.

✔ **Chemical components:** Some organisms produce unique chemicals. The ability to make certain chemicals requires specialized enzymes. These enzymes, and consequently chemicals, being present in some organisms but not in others suggests that the organisms are evolutionarily related.

✔ **Genetic information:** An organism's genetic code determines its traits, so by reading the genetic code in DNA, biologists can go right to the source of differences among species. Biologists often compare sequences of the genes for ribosomal RNA because these genes haven't changed much over evolutionary time (see Figure 9-1). Besides exact sequences, biologists also compare the *G-C ratio*, which is the percentage of DNA that's made of G-C base pairs (as opposed to A-T base pairs; see Chapter 2 for more on DNA).

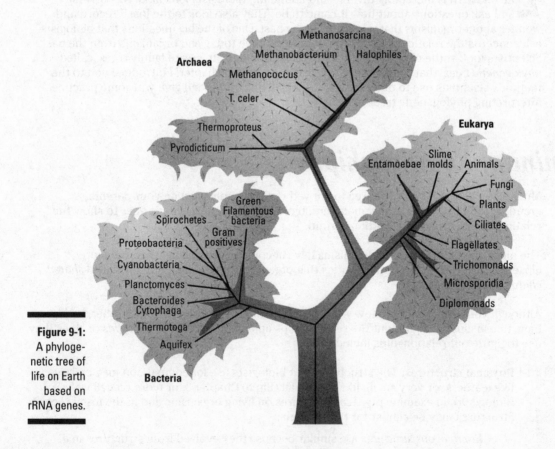

Figure 9-1: A phylogenetic tree of life on Earth based on rRNA genes.

1. Examine the following chart of characteristics for three different organisms. Which two organisms are most closely related to each other based on the number of shared characteristics?

	Organism A	Organism B	Organism C
Cells have a cell wall?	Yes	Yes	No
Cells have a nucleus?	No	Yes	Yes
G-C ratio in DNA	72%	36%	25%

Classifying Life

Figure 9-1 (see the preceding section) represents biologists' current understanding of the tree of life — the phylogenetic tree that shows relationships among all life on Earth. Each branch on the family tree leads to a new category of organisms.

The categories that biologists use for grouping related organisms are called *taxa* (singular: *taxon*).

Biologists organize all these categories into a *taxonomic hierarchy,* a naming system that ranks organisms by their evolutionary relationships. Within this hierarchy, living things are organized from the largest, most-inclusive group (called *domains*) down to the smallest, least-inclusive group (called *species*).

From largest to smallest groups, the taxonomic hierarchy is

- ✔ **Domain:** Domains group organisms by fundamental characteristics like cell structure and chemistry. The three domains are Bacteria, Archaea, and Eukarya.

- ✔ **Kingdom:** Kingdoms classify organisms based on developmental characteristics and nutritional strategy. The most familiar kingdoms are those in domain Eukarya: Animalia, Plantae, Fungi, and Protista. (Kingdoms aren't well defined for the other domains.)

- ✔ **Phylum:** Phyla separate organisms based on key characteristics that define the major groups within the kingdom.

- ✔ **Class:** Classes categorize organisms based on key characteristics that define the major groups within the phylum.

- ✔ **Order:** Orders group organisms based on key characteristics that define the major groups within the class.

- ✔ **Family:** Families classify organisms based on key characteristics that define the major groups within the order.

- ✔ **Genus:** Genera separate organisms based on key characteristics that define the major groups within the family. Genera names are typically capitalized and italicized.

- ✔ **Species:** Species categorize eukaryotic organisms based on whether they can successfully reproduce with each other and create viable offspring. (For prokaryotes, which reproduce asexually, species are defined by a set of common characteristics such as metabolic pathways and genetic similarity.) Species names are italicized.

You can think of all the categories of the taxonomic hierarchy as a set of nesting dolls like the Russian matryoshka dolls, where you open a big doll to find a smaller doll inside, which opens to reveal an even smaller doll, and so on. The only difference is that when you open the big domain doll, you find several kingdom dolls inside, and you find several phylum dolls inside the kingdom dolls, and so on.

To remember the taxonomic hierarchy, try memorizing a sentence like this one: _Dumb Kids Playing Chase On Freeways Get Squished._ The first letter of each word in the sentence represents the first letter of a category in the taxonomic hierarchy, in order. If you don't like this sentence, search the Internet for "taxonomic hierarchy mnemonic" and you'll find many more.

2.–9. Questions 2 through 9 give the complete taxonomic identification for a ball python in order from most-inclusive category to least-inclusive category. Use the following terms to identify each category.

    a. Domain

    b. Family

    c. Species

    d. Genus

    e. Order

    f. Class

    g. Kingdom

    h. Phylum

2. Eukarya

3. Animalia

4. Chordata

5. Reptilia

6. Squamata

7. Boidae

8. _Python_

9. _regius_

Figuring Out Relationships from Phylogenetic Trees

You can interpret the degree of relationship between two organisms by looking at their positions on a phylogenetic tree.

Phylogenetic trees not only show how closely related organisms are but also help map out the evolutionary history, or *phylogeny,* of life on Earth.

Based on structural, cellular, biochemical, and genetic characteristics, biologists classify life on Earth into groups that reflect the planet's evolutionary history. Just like your family began a long time ago with your original human ancestors, scientists believe that all life on Earth began from one original *universal ancestor* after the Earth formed 4.5 billion years ago. Most phylogenetic trees reflect this idea by being *rooted,* meaning they're drawn with a branch that represents the common ancestor of all the groups on the tree. In Figure 9-1, the unlabeled branch at the bottom of the tree represents the common ancestor for all organisms on the tree, which in this case is the universal ancestor of all life on Earth.

To read a phylogenetic tree like the one in Figure 9-2, look for the following information:

- The tips of the branches represent the species or other taxa that scientists compare.
- Branches meet at points called *nodes* that represent the common ancestor of the two taxa.
- Scientists call groups that branch out from the same common ancestor *sister groups*.
- An ancestor plus all its descendants form a *clade*.
- Scientists call groups that branch from the tree's base and are separate from the other groups *outgroups*. Scientists often deliberately include observations about a group that isn't very closely related to the group they're studying in order to give a tree an outgroup. A computer program that includes an outgroup helps give the tree scale by showing the group scientists are studying in relation to the larger picture of other kinds of life on Earth.

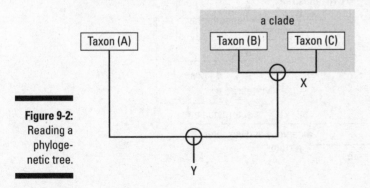

Figure 9-2:
Reading a phylogenetic tree.

Q. Examine the phylogenetic tree in Figure 9-2 and answer the following:

a. Which group is a sister group to taxon B?

b. Which group is an outgroup in this tree?

c. What represents the common ancestor to B and C?

d. What represents the common ancestor to A, B, and C?

e. What would you include to identify the clade of A, B, and C?

A. The answers to the questions are

a. The sister group to taxon B is taxon C. You can tell because they share a common ancestor.

b. In this tree, taxon A is the outgroup because it branches from the tree's base.

c. The node marked *X* represents the common ancestor to B and C. It's the connecting point for their two branches.

d. The node marked *Z* represents the common ancestor to A, B, and C. It's the connecting point for all three branches.

e. You'd include the node marked *Z*, plus taxon A, B, and C.

10.–15. Use the information in Figure 9-3 to answer the following questions.

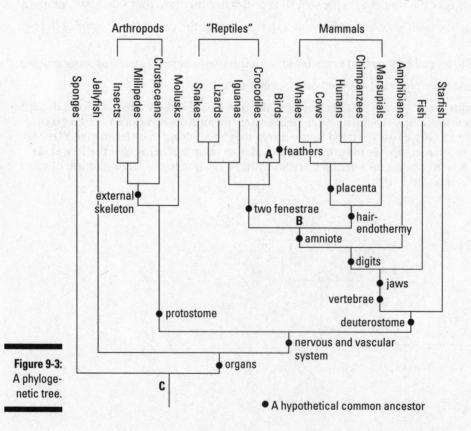

Figure 9-3:
A phyloge-
netic tree.

10. What's the sister group to cows?

11. At the top of the tree, a bracket marks the groups that are considered to belong to the reptiles. Would you consider the reptile group, as labeled, to be a true clade? If yes, why? If no, why not?

12. What represents the common ancestor of reptiles and mammals?

13. What group was used as an outgroup for this tree?

14. What does *A* represent?

15. If you created a clade that contained organisms with placentas, what other group would you need to include?

Answers to Questions on Classification and Phylogeny

The following are answers to the practice questions presented in this chapter.

1 The answer is that **organisms B and C are most closely related.**

Although organism C doesn't have a cell wall, it matches organism B for two out of three characteristics. Organisms A and B only match for one characteristic, and organisms A and C don't match at all.

2 The answer is **a. Domain.**

3 The answer is **g. Kingdom.**

4 The answer is **h. Phylum.**

5 The answer is **f. Class.**

6 The answer is **e. Order.**

7 The answer is **b. Family.**

8 The answer is **d. Genus.**

9 The answer is **c. Species.**

10 The answer is **whales** because they share a common ancestor.

11 The answer is **no.**

The reptile group doesn't include all the descendants of the common ancestor. You'd have to add birds to make it a true clade.

12 The answer is **node B.**

13 The answer is **sponges** because they branch from the base of the tree.

14 The answer is **the common ancestor for birds and crocodiles.**

15 The answer is **marsupials** because they share a common ancestor with the placental mammals.

Chapter 10

Connecting Organisms in Ecosystems

. . . .

In This Chapter

▶ Defining ecosystems, communities, and populations

▶ Understanding population dynamics

▶ Examining interactions among different species

▶ Looking at the food chain and the niches of organisms

▶ Tracing energy's flow

▶ Following the carbon cycle

. . . .

*E*cology is the branch of biology that studies how organisms connect to their environment and each other. One of the fundamental rules of ecology is that everything is connected to everything else. Organisms interact with one another and their environment as they seek the energy, matter, water, and space they need to survive. An organism's success at simply staying alive fluctuates over time, and the population of a particular group of organisms increases or decreases as conditions change. In this chapter, I present some of the fundamental characteristics of populations and show you how populations and individuals interconnect.

Ecosystems: Bringing It All Together

Life thrives in every environment on Earth, and each of those environments is its own *ecosystem,* essentially a little machine made up of living and nonliving parts that interact with one another in a particular environment. The living parts, called *biotic factors,* are all the organisms that live in the area. The nonliving parts, called *abiotic factors,* are the nonliving things in the area.

All the living things together in an ecosystem form a *community.* Within a community, each group of the same kind of organism that lives in the same area at the same time is a *population.* Figure 10-1 shows how these levels of organization intersect with one another and with other levels of biological organization.

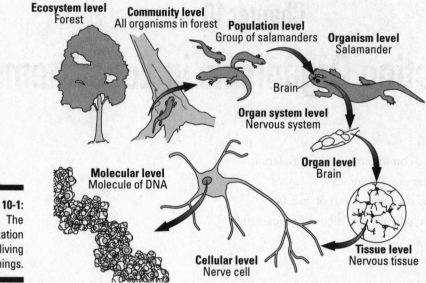

Figure 10-1:
The
organization
of living
things.

Try these questions to see how well you understand basic ecosystem terminology.

EXAMPLE

0. Imagine your backyard or a natural area near your home. Identify the abiotic and biotic factors that exist in this ecosystem.

A. Your answer depends on the environment you chose, but in general, abiotic factors include sun, rocks, air, and water. Biotic factors include all organisms such as plants, insects, birds, and mammals. Materials such as soil or a pond, which contain both living and nonliving components, can be hard to classify. In those cases, identify what parts are living or nonliving. For example, the minerals that make up soil are abiotic, while the bacteria and fungi that live in soil are biotic.

1.–3. Use the terms that follow to identify which level of biological organization is represented by each scenario.

   a. Ecosystem

   b. Community

   c. Population

1. Your friend is a beekeeper and has a hive of bees in his backyard. What does the group of bees living in the hive represent?

2. You go to the beach on a sunny day and walk on the rocks at the edge of the ocean. You see mussels and algae growing on the rocks, and you find small crabs, anemones, and fish in some tide pools. What does this rocky area represent?

3. In a forest, trees are often inhabited by birds, squirrels, insects, and fungi. What do the living things in the tree represent?

Describing Populations

Population ecology is the branch of ecology that studies the structures of populations and how they change. The unique thing about population ecologists is that they study the relationships within ecosystems by studying the properties of populations rather than individuals:

✔ **Population size** is the total number of individuals in the population.

✔ **Population density** refers to how many organisms occupy a specific area.

✔ **Dispersion** describes the distribution of a population throughout a certain area. Populations disperse in three main ways:

- **Clumped dispersion:** Organisms form clusters, with few in between.

- **Uniform dispersion:** Organisms spread evenly throughout an area.

- **Random dispersion:** Organisms scatter randomly throughout an area so that one place in the area is as good as any other for finding the organism.

✔ **Age structure** refers to the distribution of organisms of different ages in the population. **Age-structure diagrams,** like the one in Figure 10-2, show the number of individuals in each age group in a population at a particular time.

To assess the properties of populations, ecologists conduct surveys. Depending on the size and mobility of the organisms, ecologists use the following survey methods:

✔ **Total counts** count every member of the population.

✔ **Sampling methods** examine small samples of the population as representatives of the larger population. Two commonly used methods of sampling are

- **Quadrat method:** Ecologists mark off small areas of known size within a larger area (usually, they place the quadrats randomly within the larger area) and then survey the organisms within the quadrat. They use an average of the information from all the quadrats to represent the larger population.

- **Capture, mark, and release:** Ecologists capture and tag a sample of individuals from the larger population. Then, a short time later, they capture another random sample and count how many tagged individuals are in the second capture. The proportion of marked individuals in the second capture equals the proportion of the total number of tagged individuals (from the first capture) to the number of individuals in the entire population.

Q. You are an ecologist studying the recovery of California condors in the area around the Grand Canyon. You want to conduct a survey to determine the current population size. You catch 7 condors, tag them, and release them. Two months later, you catch 10 condors. Only 1 condor of your second catch is tagged. What is your estimate of the condor population? What sampling method are you using?

A. Your second catch contains 1 tagged condor out of 10, or 1/10, which equals 0.1. The proportion of tagged condors in the second catch equals the proportion of all tagged condors to the entire population. You originally tagged 7 condors, so 7 condors/X total condors = 0.1 (X is the unknown number of total condors). If you solve for total condors by multiplying both sides of the equation by X and then dividing both sides of the equation by 0.1, you get 7/0.1 = 70. The total number of condors in the area around the Grand Canyon as determined by the capture, mark, and release method is 70.

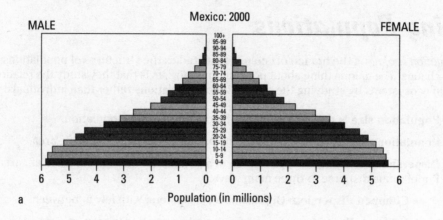

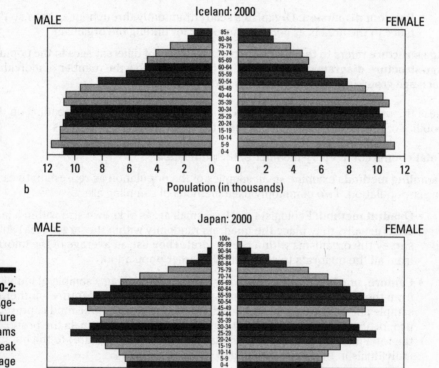

Questions 4 through 6 refer to the following scenario:

You're an ecologist who wants to estimate the size of the bull trout population in the Flathead Basin in Montana. You catch 10 bull trout, tag them, and release them. One month later, you catch 300 fish, of which only 1 is a tagged bull trout.

4. What's your estimate of the total population size of bull trout in the Flathead Basin?

    a. 1 bull trout

    b. 300 bull trout

    c. 3,000 bull trout

    d. 30,000 bull trout

5. What's the name of the method you used to estimate the bull trout population?

    a. Total count

    b. Quadrat method

    c. Capture, mark, and release

6. Why is the method you used a good choice for this particular population? In your explanation, be sure to mention why this method is a better choice than the other two methods.

Tracking Changes in Populations

Population dynamics are changes in population density over time or in a particular area. Primarily, populations increase because of births (natality) and decrease because of deaths (mortality).

The rate at which a population increases (r) depends on the relative number of births (B) and deaths (D) during a particular time interval:

$$r = B - D$$

To figure out how many individuals are added to or subtracted from a population (in other words, how much a population grows or decreases), you also have to take into account the size of the population itself. The rate of growth (G) of a population is equal to its rate of increase (r) times the population size (N):

$$G = r \times N$$

If a population has unlimited resources like food, water, and space, it has the potential to keep increasing at a steady rate, resulting in *exponential growth* like that shown in Figure 10-3a.

The maximum growth rate of a population under ideal conditions is referred to as *biotic potential*.

In nature, resources are usually limited, and population growth can be limited by a number of environmental factors, which population ecologists group into two categories:

✔ **Density-dependent factors** are more likely to limit growth as population density increases.

✔ **Density-independent factors** limit growth but aren't affected by population density.

Thus, most natural populations exhibit *logistic growth,* where the growth rate decreases as the population size increases, like that shown in Figure 10-3b.

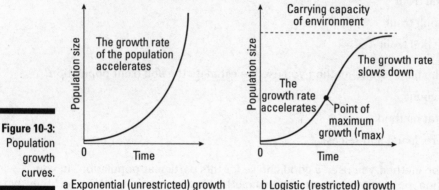

Figure 10-3:
Population growth curves.

a Exponential (unrestricted) growth   b Logistic (restricted) growth

Ecologists call the maximum population a particular habitat can sustain its *carrying capacity* (refer to Figure 10-3b). As populations approach the carrying capacity, density-dependent factors have a greater effect, and population growth slows dramatically.

Scientists have followed groups of organisms all born at the same time and looked at their *survivorship* — the number of organisms in the group that were still alive at different times after birth. They noticed that three different patterns appeared when they plotted out *survivorship curves* — graphs that plot survival over time after birth, like the ones in Figure 10-4:

- ✔ **Type I survivorship:** Most offspring survive, and organisms live out most of their life span, dying in old age.

- ✔ **Type II survivorship:** Death occurs randomly throughout the life span, usually due to predation or disease.

- ✔ **Type III survivorship:** Most organisms die young, and few members of the population survive to reproductive age. However, individuals that do survive to reproductive age often live out the rest of their life span and die in old age.

Test your understanding of population dynamics with the following questions.

Q. You put 10,000 bacterial cells in a test tube containing fresh food. These bacteria reproduce once per hour, and no significant amount of deaths occur for 12 hours. How many bacteria are in the population after just 30 minutes?

A. The starting population size (N) = 10,000. The rate of growth (r) is 50 percent every hour. So G = 0.5 × 10,000 = 5,000. After an hour, the total number in the population equals the original population plus the growth: N = 10,000 + 5,000 = 15,000.

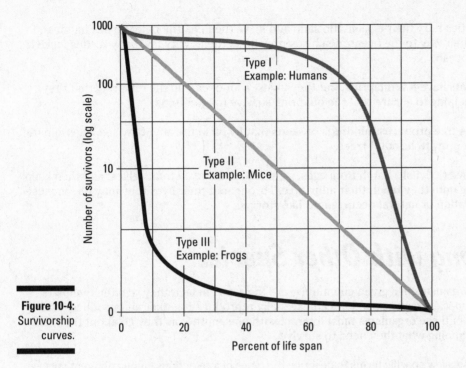

Figure 10-4:
Survivorship
curves.

7. A population of fruit flies is growing exponentially on some spoiled fruit. The intrinsic rate of increase for the population is 6 percent per day. If you start out with 10 fruit flies, how many fruit flies will you have in two days?

8.–11. Use the following terms to identify which factors affect the populations in each question.

　　a. Density-dependent

　　b. Density-independent

8. The availability of light in a forest limits the growth of plants.

9. A hurricane destroys a swamp ecosystem.

10. Birds compete for nesting sites in a forest.

11. A viral infection kills bees.

12.–15. Use the following terms to identify which type of survivorship the population in each question exhibits.

　　a. Type I

　　b. Type II

　　c. Type III

12. Sea turtles bury their eggs in the sand and leave them. As the baby turtles hatch and make their way to the ocean, sea birds prey upon them. Very few baby turtles make it to the ocean.

13. Elephants have few offspring, but they invest a lot of care in the offspring that they have, helping to ensure that the offspring survive to adulthood.

14. A maple tree produces hundreds of seeds that fly off in the air. Many seeds germinate, but few grow to be adult trees.

15. Sea jellies (jellyfish) hatch from eggs, grow into small planktonic jellies, and then keep growing until they reach their adult size. Throughout their lives they may die because of predation or natural occurrences like storms.

Getting Along with Other Species

Not all the organisms in a given community are the same. In fact, they're often members of different *species* (organisms that can't sexually reproduce together and produce fertile offspring). Yet these organisms must interact with one another as they go about their daily business of finding what they need to survive.

Ecologists use a few specific terms to describe the types of interactions among different species:

- **Predation:** One organism *(predator)* eats another *(prey)*.

- **Competition:** Both organisms suffer as they compete with each other for limited resources such as food, water, or space.

- **Symbiosis:** Two organisms live together for a large part of their life cycle. Three types of symbiosis may occur:

   • **Parasitism:** One organism benefits at the expense of the other.

   • **Mutualism:** Both organisms benefit.

   • **Commensalism:** One organism benefits but the other isn't benefitted or harmed.

16.–21. Use the following terms to identify the type of relationship described in each question.

   a. Predation

   b. Competition

   c. Parasitism

   d. Mutualism

   e. Commensalism

16. Bacteria live in cows' stomachs, where they break down the cellulose in the grass that cows eat. The bacteria get food from the cellulose, and the cows get food from the bacteria.

17. A group of tomato plants planted close together in a garden shade one another with their leaves.

18. Barnacles grow on the sides of whales. As whales swim through the ocean, the barnacles are carried through nutrient-rich water. The barnacles don't hurt or help the whales.

19. A dog remains very thin despite eating plenty of food because it has a tapeworm living in its intestines. The tapeworm receives nutrients from the dog's food.

20. A human kills and eats a chicken.

21. *E. coli* lives in the intestines of humans. The bacterium gets nutrients from the food the humans eat. Humans get a vitamin from the bacterium that they don't get from their food.

Discovering the Job Descriptions of Organisms

Organisms interact with their environment and with other organisms to acquire energy and matter for growth. The interactions among organisms influence behavior and help the organisms establish complex relationships. Each organism occupies a certain *niche* that includes both the environment where it lives and the role it plays in that environment. Many biologists refer to an organism's niche as its *job description* in its environment.

One of the most fundamental components of an organism's niche is defined by how the organism gets its food. Based on this characteristic, all the various organisms in an ecosystem can be divided into four categories, called *trophic levels:*

- ✔ **Producers** make their own food by processes like photosynthesis. Producers can also be called *autotrophs* (see Chapter 4 for more on autotrophs and the process of photosynthesis).

- ✔ **Primary consumers** eat producers. Because producers are mainly plants, primary consumers are also called *herbivores* (plant-eating animals).

- ✔ **Secondary consumers** eat primary consumers. Because primary consumers are animals, secondary consumers are also called *carnivores* (meat-eating animals).

- ✔ **Tertiary consumers** eat secondary consumers, so they're also carnivores.

A special category is reserved for organisms that eat the dead because they feed off all the other trophic levels. These *decomposers,* like bacteria and fungi, release enzymes onto dead organisms, breaking them down into smaller components which they then absorb.

Organisms in the different trophic levels are linked together in a *food chain,* a sequence of organisms in a community in which each organism feeds on the one below it in the chain. When all the food chains from an ecosystem are put together, they form an interconnected *food web.*

22.–25. Use the following terms to label the food chain in Figure 10-5.

    a. Primary consumer

    b. Producer

    c. Decomposer

    d. Secondary consumer

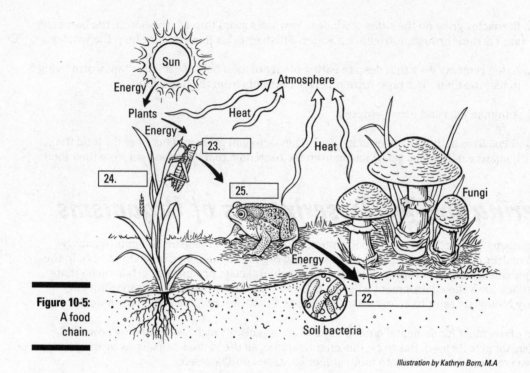

Figure 10-5:
A food
chain.

Illustration by Kathryn Born, M.A

Following the Flow of Energy through Ecosystems

The energy living things need to grow flows from the sun to plants and then from one organism to another through food. As energy flows through ecosystems, it transfers from one place to another and transforms from one type of energy to another (for more on how energy flows, check out Chapter 4).

Chapter 4 introduces the *first law of thermodynamics* which says that energy can't be created or destroyed. The consequence of this law is that every living thing has to get its energy from somewhere. Even producers, who make their own food, can't make their own energy — they capture energy from the environment and store it in the food they make.

Another important law that governs how energy moves through ecosystems is the *second law of thermodynamics*. Because of this law, when energy is transferred in living systems, some of the energy is transformed into heat energy. The impact of this law on ecosystems is that no energy transfer is 100 percent efficient. After energy is transferred to heat, it's no longer useful as a source of energy to living things.

Producers capture about 1 percent of the energy that's available from the sun. As energy moves through food webs, each trophic level can capture only about 10 percent of the energy that was available to the trophic level it feeds on. Ecologists call this rule of thumb the *ten percent rule*. Ecologists illustrate the flow of energy through trophic levels by creating *energy pyramids* (also called *trophic pyramids;* see Figure 10-6).

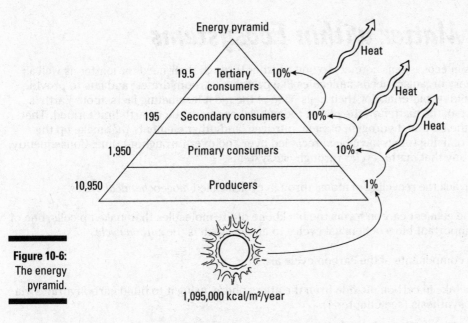

Figure 10-6:
The energy
pyramid.

Energy pyramid

19.5	Tertiary consumers	10%
195	Secondary consumers	10%
1,950	Primary consumers	10%
10,950	Producers	1%

Heat

Heat

Heat

1,095,000 kcal/m²/year

26.–30. Use the following terms to fill in the blanks and describe the flow of energy in each question.

    a. Transferred

    b. Transformed

    c. Source

    d. Receiver

26. During photosynthesis, plants capture energy from the sun and store it in carbohydrates. Thus, light energy is _____ into chemical energy.

27. During photosynthesis, the sun is the _____ of the energy.

28. During photosynthesis, the plant is the _____ of the energy.

29. As people exercise, they use stored chemical energy to move their muscles. Thus, chemical energy is _____ into motion energy.

30. As people exercise, some of their stored chemical energy is converted to heat. This heat passes through their bodies and into the surrounding air. Thus, heat energy is _____ to the environment.

31. Explain why top predators in an ecosystem rarely go beyond tertiary consumers.

Recycling Matter within Ecosystems

Organisms in ecosystems connect to one another through their need for matter as well as energy. Every organism needs molecules like proteins, carbohydrates, and fats to provide the raw building materials for their cells. One of the most fascinating facts about Earth is that almost all the matter on the planet today has been here since Earth first formed. That means all the carbon, hydrogen, oxygen, nitrogen, and other elements that make up the molecules of living things have been recycled over and over throughout time. Consequently, ecologists say that matter *cycles* through ecosystems.

Scientists track the recycling of atoms through cycles called *biogeochemical cycles*.

Because the element carbon forms the backbone of the molecules that make up cells, one of the most important biogeochemical cycles to life on Earth is the *carbon cycle*.

The major components of the carbon cycle are

- ✔ Plants take in carbon dioxide from the atmosphere, using it to build carbohydrates via photosynthesis (see Chapter 4).

- ✔ Carbon moves through food chains as organisms eat other organisms, incorporating the carbon-containing molecules from their food into the organic molecules that make up their own bodies.

- ✔ Carbon moves from living things back to the environment as all types of organisms use some of their food molecules as a source of energy. To transfer energy from their food to their cells, organisms break down the food molecules into carbon dioxide and water in the process of cellular respiration (see Chapter 4).

- ✔ In addition to being animals that do cellular respiration, humans also contribute to the carbon cycle through industrial processes. The combustion of fossil fuels produces carbon dioxide and water, contributing to the amount of carbon dioxide in the atmosphere.

- ✔ The fossil fuels that humans use today formed hundreds of millions of years ago when dead plants were deposited in the Earth in a way that slowed decomposition, allowing the organic molecules to be converted by geologic forces into oil, coal, and natural gas.

- ✔ Carbon dioxide diffuses from the air into the oceans, where it exists in a dissolved form, creating another large reservoir of carbon.

Q. What parts of the carbon cycle contribute carbon dioxide to the atmosphere?

A. Carbon dioxide returns to the atmosphere through cellular respiration by all types of living things and by combustion of wood and fossil fuels.

32.–40. Use the terms that follow to label the events of the carbon cycle shown in Figure 10-7.

   a. Cellular respiration by animals

   b. Cellular respiration by plants

   c. Cellular respiration by decomposers (decomposition)

   d. Atmospheric carbon dioxide

   e. Photosynthesis

   f. Dissolved carbon dioxide

   g. Organic carbon in living things

   h. Combustion of fossil fuels

   i. Conversion of organic carbon to fossil fuels

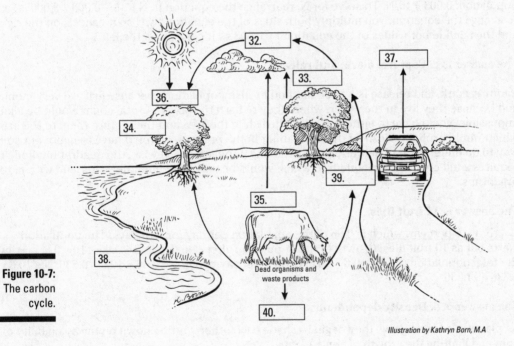

Figure 10-7:
The carbon
cycle.

Illustration by Kathryn Born, M.A

Answers to Questions on Ecosystems

The following are answers to the practice questions presented in this chapter.

1 The answer is **c. Population.**

2 The answer is **a. Ecosystem.**

The rocky intertidal zone contains interconnected biotic and abiotic factors.

3 The answer is **b. Community.**

4 The answer is **c. 3,000 bull trout.**

In your second catch, only 1 of 300 fish is a tagged bull trout, which is a proportion of $1 \div 300$, or 0.003. That means your original catch of 10 bull trout should represent 0.003 of the total population: $0.003 = 10/N$. To solve for N, rearrange the equation to $N = 10 \div 0.003 = 3,000$. (To rearrange the equation, you multiply both sides of the equation by N so it cancels on the right and then divide both sides of the equation by 0.003 so it cancels on the left.)

5 The answer is **c. Capture, mark, and release.**

6 Capture, mark, and release is a good method for fish populations because fish are very mobile and because they live in the water, where they're hard to survey. A total count would be almost impossible; you'd have to either catch all the fish in the area and then return them to the basin, which would probably destroy the fish habitat in the process, or you'd have to figure out some way to do underwater observations on a moving population! Likewise, the quadrat method still requires a full count of each quadrat, which would be very difficult underwater and with mobile organisms.

7 The answer is **11 fruit flies.**

$G = rN$; r = 6 percent, which is 0.06 of the population per day for two days. The population starts out as 10 fruit flies. Growth of the population after one day is $G = 0.06 \times 10 = 0.6$, making the total population $N = 10 + 0.6 = 10.6$. Growth on day two is $G = 0.06 \times 10.6 = 0.636$; $N = 10.6 + 0.636 = 11.236$.

8 The answer is **a. Density-dependent.**

As plants in a forest grow, they begin to shade each other, cutting down on the availability of light and limiting the growth of some plants.

9 The answer is **b. Density-independent.**

10 The answer is **a. Density-dependent.**

11 The answer is **a. Density-dependent.**

12 The answer is **c. Type III.**

13 The answer is **a. Type I.**

14 The answer is **c. Type III.**

15 The answer is **b. Type II.**

16 The answer is **d. Mutualism.**

17 The answer is **b. Competition** (for light).

18 The answer is **e. Commensalism.**

19 The answer is **c. Parasitism.**

The worm is the parasite.

20 The answer is **a. Predation.**

Humans are the predator, and chickens are the prey.

21 The answer is **d. Mutualism.**

Other strains of *E. coli* that occasionally infect humans show parasitism.

22 – 25 The following is how Figure 10-5 should be labeled:

22 **c. Decomposer;** 23 **a. Primary consumer;** 24 **b. Producer;** 25 **d. Secondary consumer.**

26 The answer is **b. Transformed.**

27 The answer is **c. Source.**

28 The answer is **d. Receiver.**

29 The answer is **b. Transformed.**

30 The answer is **a. Transferred.**

31 Each trophic level captures energy and transfers it for its own growth, maintenance, and reproduction. Because every energy transfer includes some energy transfer to heat, which passes
into the environment, only a small percentage of the total energy an organism captures is
stored in its body. So when one trophic level feeds on another, only about 10 percent of the
energy captured by the lower level is available to the upper level. So much energy is transferred to the environment along the way that there isn't enough available energy to support
many trophic levels.

32 – 40 The following is how Figure 10-7 should be labeled:

32 **d. Atmospheric carbon dioxide;** 33 **b. Cellular respiration by plants;** 34 **g. Organic carbon
in living things;** 35 **a. Cellular respiration by animals;** 36 **e. Photosynthesis;** 37 **h. Combustion
of fossil fuels;** 38 **f. Dissolved carbon dioxide;** 39 **c. Cellular respiration by decomposers
(decomposition);** 40 **i. Conversion of organic carbon to fossil fuels.**

Chapter 11

Evaluating the Effects of Evolution

*U*nderstanding evolution is essential for understanding how organisms change over time. In this chapter, I present you with the fundamental explanation of evolution first proposed by Charles Darwin and show you the many lines of scientific research that support this theory. You get a chance to think about how evolution happens and practice your skills in identifying the key factors that influence this process.

Defining Evolution

Biological evolution refers to the change of living things over time. Charles Darwin introduced the world to this concept in his 1859 work, *On the Origin of Species.* In this book, Darwin proposed that living things descend from their ancestors but that they can change over time.

Darwin concluded that biological evolution occurs as a result of *natural selection,* which is the theory that, in any given generation, some individuals are more likely to survive and reproduce than others.

The theory of natural selection is often referred to as "survival of the fittest." In biological terms, fitness has nothing to do with how often you pump iron or go to spin class. Biological fitness is basically your ability to produce offspring. So survival of the fittest really refers to the passing on of those traits that enable individuals to survive and successfully reproduce.

Figure 11-1 illustrates natural selection in action. If a visual predator, such as an eagle, is cruising for its lunch, it will most likely eat the animals it can see most easily. If the eagle's prey is mice, which can be white or dark (see Figure 11-1a), and the mice live in the forest on dark-colored soil, the eagle will see the white mice more easily. Over time, if the eagles in the area keep eating more white mice than dark mice (see Figure 11-1b), more dark mice will reproduce. Dark mice have genes that specify dark-colored fur, so their offspring will also have dark fur. If the eagles continue to prey upon mice in the area, the population of mice in the forest will gradually begin to have more dark-colored individuals than white individuals (see Figure 11-1c).

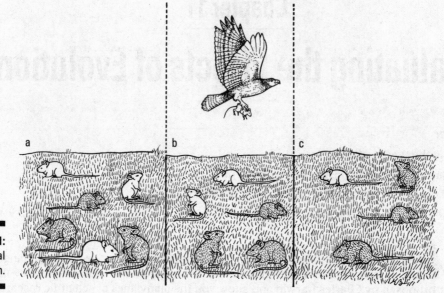

Figure 11-1:
Natural
selection.

Illustration by Kathryn Born, M.A

In this example, the eagle is the *selection pressure* — an environmental factor that causes some organisms to survive and others not to survive.

For natural selection to occur in a population, several conditions must be met:

✔ Individuals in the population must produce more offspring than can survive.

✔ Those individuals must have different characteristics.

✔ Some characteristics must be passed on from parents to offspring.

✔ Selective pressure favors organisms with the best-suited characteristics for their environment.

If these four conditions are met, the new generation of individuals will be different from the original generation in the frequency and distribution of traits, which is pretty much the definition of biological evolution.

Questions 1 through 3 refer to the following story:

A classic example of how natural selection can change a population occurred in Britain during the Industrial Revolution. Peppered moths in England are eaten by birds, which hunt by sight. Prior to the industrialization of England, a light-colored form of the peppered moth was more abundant, even near cities. Immediately following the rise in industry, which spread coal soot over urban areas, a dark-colored form of the peppered moth became more abundant in urban areas, while light-colored moths remained more abundant in the countryside. After air-pollution reforms took effect in England and the coal dust was gone, the light-colored form of the moth became more abundant in urban areas.

1. What was the selection pressure acting on the moths?

2. How did the definition of fitness change for the moths in urban areas before and after the Industrial Revolution?

3. How would the story be different if wing color in moths wasn't an inherited characteristic (for example, if wing color were determined randomly at birth based on environmental temperature)?

Predicting the Outcome of Natural Selection

Natural selection may cause populations to change, but exactly how they change depends on the specific selective pressures at a given time. Individuals within a population may evolve to be more similar to or more different from one another depending on the specific circumstances and selection pressures.

The two most extreme outcomes of natural selection are extinction and speciation. Species that can't adapt to changing environmental conditions may become *extinct,* or disappear from Earth. On the other extreme, new species may arise when a population accumulates so many changes that it can no longer mate with related organisms. Biologists call the creation of new species *speciation.*

Four types of natural selection may act to cause changes in populations:

- **Stabilizing selection:** This type eliminates extreme or unusual traits. Individuals with the most common traits are considered best adapted, which maintains the frequency of common traits in the population. Over time, nature selects against extreme variations of the trait.

- **Directional selection:** Traits at one end of a spectrum of traits are selected for, whereas traits at the other end of the spectrum are selected against. Over generations, the selected traits become common and the other traits become rarer and rarer until they're eventually phased out.

- **Disruptive selection:** The environment favors extreme or unusual traits and selects against the common traits. Over time the traits at opposite ends of the trait spectrum dominate.

- **Sexual selection:** Females increase the fitness of their offspring by choosing males with superior fitness; females are therefore concerned with quality. Males contribute most to a species' fitness by maximizing the quantity of offspring they produce. Competition among males for opportunities to mate exists in the form of strength contests, and traits that give a male an advantage in a strength contest evolve. Because females choose their mates, males also develop traits to attract females, such as certain mating behaviors or bright coloring.

Biological evolution happens to populations, not individuals. Individuals live or die and reproduce or don't reproduce depending on their circumstances. Individuals themselves can't evolve in response to a selection pressure, but over time, a species *can* evolve.

Imagine a giraffe whose neck isn't quite long enough to reach the tastiest leaves at the top of the tree. That individual giraffe can't suddenly grow its neck longer to reach the leaves. However, if another giraffe in the herd has a longer neck, gets more leaves, grows better, and makes more calves that inherit his long neck, then future generations of giraffes in that area may have longer necks.

4.–7. Use the terms that follow to identify the type of natural selection that may have produced each characteristic.

    a. Stabilizing selection

    b. Directional selection

    c. Disruptive selection

    d. Sexual selection

4. Female dung beetles prefer to mate with larger male dung beetles.

5. Frequent lawn mowing leads to faster-growing weeds.

6. Some humans are very tall, and some humans are very short, but most humans are somewhere in between.

7. A few finches landed on the Galapagos Islands, which had no birds living there. From those original finches, many different species of finches evolved, each one having a different type of beak and specializing in a different type of food.

Supporting the Theory of Evolution

Since Darwin first proposed his ideas about biological evolution and natural selection, different lines of research from many different branches of science have produced evidence supporting his belief that biological evolution occurs in part because of natural selection.

Because a great amount of data supports the idea of biological evolution through natural selection, and because no scientific evidence has yet been found to prove this idea false, this idea is considered a *scientific theory.* (For more on the importance of theories in science, see Chapter 1.)

Here's a brief summary of the evidence that supports the theory of evolution by natural selection:

    ✔ **Biochemistry** is the study of the basic chemistry and processes that occur in cells. The biochemistry of all living things on Earth is incredibly similar, showing that all of Earth's organisms share a common ancestry.

    ✔ **Comparative anatomy** is the comparison of the structures of different living things. Figure 11-2 compares the skeletons of humans, cats, whales, and bats, illustrating how similar they are even though these animals live unique lifestyles in very different environments. The best explanation for similarities like the ones among these skeletons is that the various species on Earth evolved from common ancestors.

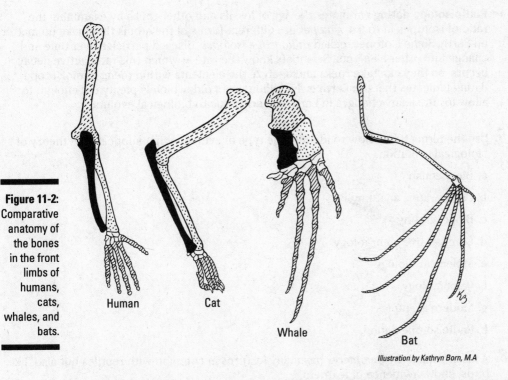

Figure 11-2: Comparative anatomy of the bones in the front limbs of humans, cats, whales, and bats.

Human

Cat

Whale

Bat

Illustration by Kathryn Born, M.A

✔ **Biogeography,** the study of living things around the globe, helps solidify Darwin's theory of biological evolution. Basically, if evolution is real, you'd expect groups of organisms that are related to one another to be clustered near one another because related organisms come from the same common ancestor. On the other hand, if evolution isn't real, there's no reason for related groups of organisms to be found near one another. When biogeographers compare the distribution of organisms living today or those that lived in the past (from fossils), they find that species are distributed around Earth in a pattern that reflects their genetic relationships to one another.

✔ **Comparative embryology** compares the embryos of different organisms. The embryos of many animals, from fish to humans, show similarities that suggest a common ancestor.

✔ **Molecular biology** focuses on the structure and function of the molecules that make up cells. Molecular biologists have compared gene sequences among species, revealing similarities among even very different organisms.

✔ **Paleontology** is the study of prehistoric life through fossil evidence. The *fossil record* (all the fossils ever found and the information gained from them) shows detailed evidence of the changes in living things through time.

✔ **Modern examples** of biological evolution can be measured by studying the results of scientific experiments that measure evolutionary changes in the populations of organisms that are alive today. In fact, you need only look in the newspaper or hop online to see evidence of evolution in action in the form of antibiotic-resistant bacteria.

✔ **Radioisotope dating** estimates the age of fossils and other rocks by examining the ratio of isotopes in rocks. *Isotopes* are different forms of the atoms that make up matter on Earth. Some isotopes, called *radioactive isotopes,* discard particles over time and change into other elements. Scientists know the rate at which this radioactive decay occurs, so they can take rocks and analyze the elements within them. Radioisotope dating indicates that the Earth is 4.5 billion years old, which is plenty old enough to allow for the many changes in Earth's species due to biological evolution.

8.–15. Use the terms that follow to identify the type of evidence that supports the theory of biological evolution.

   a. Biochemistry

   b. Comparative anatomy

   c. Biogeography

   d. Comparative embryology

   e. Molecular biology

   f. Paleontology

   g. Modern examples

   h. Radioisotope dating

   8. A fossil named *Archaeopteryx* has many features in common with reptiles but also, like birds, shows evidence of feathers.

   9. The genetic code of all life on Earth is written in the same chemical building blocks.

   10. Some genes from the bacterium *E. coli* have sequences that are similar to genes found in humans.

   11. In the 1940s, infections by the bacterium *Staphylococcus aureus* could be treated successfully with penicillin. Today, populations exist that are completely resistant to natural penicillin, as well as to partially modified penicillins like methicillin. These populations, called MRSA, are very challenging to modern medical professionals.

   12. Whales have tiny, useless bones inside the rear portion of their bodies that are very similar to the bones found in vertebrate legs.

   13. Human embryos have gill slits like those seen in fish embryos. (Developing fish retain their gill slits, whereas humans don't.)

   14. Marsupial mammals (mammals like kangaroos that protect their young in a pouch) live in just a few places in the world today — Australia, South America, and part of North America. Although Australia is now thousands of miles away from the Americas, in the past the three continents were connected as one larger land mass.

   15. Fossils of the earliest life forms on Earth, which look like bacterial cells, occur in rocks that scientists estimate to be 3.5 billion years old.

Answers to Questions on Evolution

The following are answers to the practice questions presented in this chapter.

1 The answer is **predation by birds.**

2 Before the Industrial Revolution, highest fitness in urban areas occurred in light-colored moths. Just after the Industrial Revolution, the highest fitness occurred in dark-colored moths. After clean-air reforms led to a cleaner environment, higher fitness again occurred in light-colored moths.

3 You'd predict that the frequency of moth colors wouldn't change despite the changing landscape (assuming no temperature change was occurring). The same numbers of light- and dark-colored moths would be seen in each generation, randomly or changing in response to changes in temperature.

4 The answer is **d. Sexual selection.**

5 The answer is **b. Directional selection.**

6 The answer is **a. Stabilizing selection.**

7 The answer is **c. Disruptive selection.**

8 The answer is **f. Paleontology.**

9 The answer is **a. Biochemistry.**

10 The answer is **e. Molecular biology.**

11 The answer is **g. Modern examples.**

12 The answer is **b. Comparative anatomy.**

13 The answer is **d. Comparative embryology.**

14 The answer is **c. Biogeography.**

15 The answer is **h. Radioisotope dating.**

Part IV

Getting to Know the Human Body

The 5th Wave   By Rich Tennant

"For the last time – pregnant vegetarians do not give birth to Cabbage Patch Dolls!"

In this part . . .

Human bodies are among the most complex in the animal kingdom, with specialized organ systems that perform the functions necessary to sustain life. The muscular and skeletal systems help people stand and move, the nervous and endocrine systems send signals that lead to responses to the environment, and the respiratory and circulatory systems work together to bring oxygen to all the body's cells. In short, it's complicated. But it's also fascinating.

These examples are just a sample of what you find in this part as I introduce the fundamentals of the many organ systems in the human body. I also take a peek at some of the different ways that other animals do things.

Chapter 12

Building Bodies with the Skeletal and Muscular Systems

. . . .

In This Chapter

▶ Differentiating friction and gravity

▶ Breaking down skeleton and bone types

▶ Recognizing different kinds of joints

▶ Examining muscle structure and function

▶ Understanding the process of muscle contraction

. . . .

The coordinated efforts of muscles and skeletons make animal movements possible. Muscles and skeletons help animals resist the forces of gravity and friction so that they can stand, swim, fly, and jump. Muscles pull or push, and skeletons give the muscles something to pull or push against. In this chapter, you find out all about how animals move from place to place as you discover the different types of skeletons and the fundamentals of muscle function.

Moving Around with Friction and Gravity

Every type of *locomotion*, movement from one place to another, requires animals to use energy to overcome the forces of friction and gravity that would otherwise hold them to Earth.

▶ **Friction** is the force that pushes back on any movement of two objects in contact with each other. The force of friction due to movement through air or water is called *resistance*.

▶ **Gravity** is the force that pulls all objects that have mass toward each other.

Each animal is adapted for the environment it lives in and the type of locomotion it performs. Think about some of the animals you're familiar with and how their structure helps them move in their environment, and then answer the following questions.

1.–5. For Questions 1 through 5, answer the following: Which of the two forces (gravity or friction) does the adaptation target the most?

    a. Gravity

    b. Friction (resistance)

1. Humans have large muscles in their legs.

2. Birds have hollow bones that make their bodies light.

3. Fish are covered with a layer of slippery mucus.

4. Nematodes that burrow through the soil have thin, tubular bodies.

5. Giraffes have strong, bony skeletons.

Getting Support from Skeletons and Bones

Skeletons give muscles something to pull against; support the body's weight; store minerals like calcium and phosphorus; and produce blood cells in the bone marrow. However, not all animals have the same type of skeleton. Following are the three different kinds of skeletons you may see in your study of biology:

- ✔ **Hydrostatic skeletons** are basically chambers filled with water. Animals with this skeleton type move and change their shape by squeezing their water-filled chambers — just like what happens when you squeeze a water balloon.

- ✔ **Exoskeletons** are hard exterior coverings found on the outside of the body.

- ✔ **Endoskeletons** exist within an animal's body.

Animals with hydrostatic skeletons and exoskeletons are considered *invertebrates,* meaning they don't have a backbone. Animals with endoskeletons, like you, are considered *vertebrates* because they have a backbone. All vertebrate skeletons — whether they belong to humans, snakes, bats, or whales — developed from the same ancestral skeleton (which explains why you may notice similarities between your skeleton and that of your pet dog or cat). Today, these animals show their relationship to one another in part due to *homologous structures* — structures that are equivalent to one another in their origin.

All vertebrate skeletons have two main parts (which you can see in Figure 12-1):

- ✔ **The axial skeleton:** This part supports the animal's central column, or *axis.* It includes the skull, the backbone *(vertebral column),* and the rib cage.

- ✔ **The appendicular skeleton:** This part extends from the axial skeleton out into the arms and legs *(appendages).* It includes the shoulders, the pelvis, and the bones of the arms and legs.

You should also know the names of some of your major bones:

- ✔ Your *skull* protects your brain.

- ✔ Your *pectoral girdle* includes your collarbones *(clavicles)* and your shoulder bones *(scapulae).*

- ✔ Your *sternum* is the central bone in your chest, to which your ribs and collarbones attach.

✔ Your *vertebrae* are the small bones in your back that protect your spinal cord.

✔ Your *ribs* protect your lungs.

✔ Your *humerus* is the long bone of your upper arm.

✔ Your *ulna* is the larger of the two bones in your forearm.

✔ Your *radius* is the smaller of the two bones in your forearm.

✔ Your *carpals* are the eight small bones that form your wrist.

✔ Your *metacarpals* are the bones of your hand.

✔ Your *phalanges* are your finger bones and toe bones. In your fingers, the phalanges are arranged in pairs in your thumb and in triplets in your fingers. In your toes, the phalanges are arranged in a pair in your big toes and in triplets in your other toes.

✔ Your *pelvic girdle* includes your two hipbones and your tailbone.

✔ Your *femur,* or the long bone of your thigh, is the longest bone in your body.

✔ Your *patella* is your kneecap.

✔ Your *tibia* is your shinbone.

✔ Your *fibula* is the smaller bone in your leg that runs alongside your tibia.

✔ Your *tarsals* are the seven small bones that make up your ankle.

✔ Your *metatarsals* are the bones of your feet.

If you've ever watched an old Western movie, you've probably seen images of bones bleached white by the sun and scattered alongside a pioneer trail. These dry white bones are very different from the living bones in your body. Bone is actually a moist, living tissue that contains different layers and cell types.

✔ **Fibrous connective tissue** covers the exterior of bones and helps heal bone breaks by forming new bone.

✔ **Bone cells (osteocytes)** give cells their hard nature. The cells live in and produce a *bone matrix* made of collagen that's been hardened by the attachment of calcium and phosphate crystals.

✔ **Cartilage** covers the ends of bones and protects them from damage as they rub against one another.

The tissues found within living bone fall into two categories:

✔ **Spongy bone tissues** are filled with little holes, similar to those you see in volcanic rocks. These holes are filled with *red bone marrow,* which is the tissue that produces your blood cells.

✔ **Compact bone tissues** are hard and dense. A cavity within compact bone is filled with *yellow bone marrow,* which is mostly stored fat. If the body suddenly loses a large amount of blood, it converts the yellow bone marrow to red bone marrow so that blood cell production can be increased.

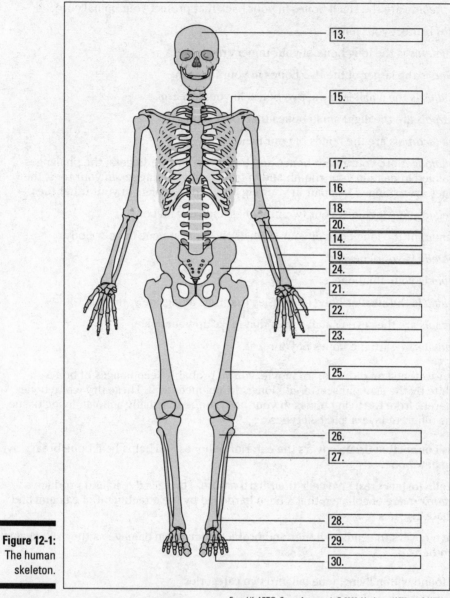

Figure 12-1:
The human
skeleton.

13.

15.

17.

16.

18.

20.

14.

19.

24.

21.

22.

23.

25.

26.

27.

28.

29.

30.

From LifeART®, Super Anatomy 1, © 2002, Lippincott Williams & Wilkins

6.–10. Use the terms that follow to identify the type of skeleton found in the animal.

 a. Hydrostatic skeleton

 b. Exoskeleton

 c. Endoskeleton

6. A dog

7. A beetle, like a ladybug

8. An earthworm

9. A shark (hint: sharks are cartilaginous fish)

10. A bird

11. State one advantage and one disadvantage for each type of skeleton.

12. Using colored pencils or a highlighter, lightly shade the axial skeleton in Figure 12-1 one color and shade the appendicular skeleton a different color.

13.–30. Use the terms that follow to label the bones in Figure 12-1.

    a. Humerus

    b. Skull

    c. Metacarpals

    d. Pelvic girdle

    e. Vertebrae

    f. Sternum

    g. Phalanges of the hand

    h. Fibula

    i. Radius

    j. Ulna

    k. Femur

    l. Tarsals

    m. Tibia

    n. Phalanges of the foot

    o. Ribs

    p. Metatarsals

    q. Pectoral girdle

    r. Carpals

31. Where would you expect to find the greatest amount of calcium?

    a. Bone matrix

    b. Cartilage

    c. Red bone marrow

    d. Yellow bone marrow

    e. Connective tissue

32. Where would you expect to find the greatest amount of fat?

    a. Bone matrix

    b. Cartilage

    c. Red bone marrow

    d. Yellow bone marrow

    e. Connective tissue

33. Which produces blood cells?

    a. Bone matrix

    b. Cartilage

    c. Red bone marrow

    d. Yellow bone marrow

    e. Connective tissue

34. Which covers the outside of bone?

    a. Bone matrix

    b. Cartilage

    c. Red bone marrow

    d. Yellow bone marrow

    e. Connective tissue

This Joint Is Jumpin'

Joints are structures where two bones are attached. *Movable* or *synovial* joints allow bones to move relative to each other. In many areas of the body, strong, fibrous connective tissues called *ligaments* stabilize joints.

Three common types of movable joints enable most of the movements of animals:

- **Ball-and-socket joints** allow movement in many directions. They consist of a bone with a rounded, ball-like end that fits into another bone, which has a smooth, cup-like surface.

- **Pivot joints** allow you to swivel a bone. They occur when one bone pivots or rotates around another bone that remains stationary.

- **Hinge joints** allow you to bring two bones close together or move them farther apart, much like you open and close a book. In hinge joints, a convex surface forms a joint with a concave surface.

35.–37. Move your body and use the following terms to figure out what kind of movable joint exists at each of these locations.

    a. Ball-and-socket joint

    b. Pivot joint

    c. Hinge joint

35. Where your skull meets the top of your spine

36. Where your humerus meets your ulna

37. Where your femur meets your pelvic girdle

Flexing Your Knowledge of Muscles

Muscle tissues are extremely important to your body — and not just because they help you look good at the pool. Muscles do many things to keep you alive and going strong:

- ✔ **Muscles allow you to stand upright.** Your muscles contract so that you can push against the surface of the Earth, defying gravity to stand upright.

- ✔ **Muscles make it possible for you to move.** Your muscles control every little movement that your body performs, from the smallest blink to the largest leap.

- ✔ **Muscles allow you to digest your food.** Muscles all along your digestive tract squeeze to keep food moving along in a process called *peristalsis*.

- ✔ **Muscles affect the rate of blood flow.** Blood vessels expand and contract using their muscle tissue, and your heart muscle contracts to move blood through your circulatory system.

- ✔ **Muscles help to maintain normal body temperature.** The chemical reactions inside muscle cells and the sliding of muscle filaments both produce heat that helps maintain your body temperature.

- ✔ **Muscles hold your skeleton together.** The ligaments and tendons at the ends of your muscles wrap around joints, holding them together.

Muscle tissues are made up of cells called *muscle fibers* (see Figure 12-2). Each muscle fiber contains many *myofibrils* — the parts of the muscle fiber that contract. The myofibrils line up right next to each other, giving muscles a striped, or *striated*, appearance. Myofibrils contract because of the sliding action of two filamentous cytoskeletal proteins, called *actin* and *myosin* (see Chapter 3 for more on the cytoskeleton):

- ✔ **Actin filaments,** or *thin filaments,* consist of two strands of actin wound around each other.

- ✔ **Myosin filaments,** or *thick filaments*, contain groups of *myosin.* Myosin filaments have bulbous ends called *myosin heads;* in muscle, multiple strands of myosin arrange with their heads pointed in opposite directions so that both ends of thick filaments look bulbous.

Thin and thick filaments are organized into repeating units called *sarcomeres* (see Figure 12-2). Dark lines called *Z-lines* mark off the boundaries of each sarcomere. Thin filaments attach to the Z-lines at both ends of the sarcomere, while thick filaments are unattached. Each myofibril contains thousands of sarcomeres.

Three types of muscle tissue exist within your body:

- ✔ **Cardiac muscle** makes up the heart. The fibers of cardiac muscle are branched, cylindrical cells that have one nucleus and striations. Cardiac muscle contraction is totally *involuntary,* meaning it occurs without nervous stimulation and doesn't require conscious control.

- ✔ **Smooth muscle** lines the walls of internal organs that are hollow, like the stomach, bladder, intestines, and lungs. The fibers of smooth muscle tissue are spindle-shaped and have one nucleus. Smooth muscle gets its name from the fact that it doesn't have horizontal striations like other muscle tissues (so it looks smooth). The fibers form sheets of tissue by lining up in parallel lines. Smooth muscle contraction occurs involuntarily and more slowly than skeletal muscle contraction, which means smooth muscle can stay contracted longer than skeletal muscle and not fatigue as easily.

✔ **Skeletal muscle** is probably what you think of when you picture a muscle. The cylindrical fibers (cells) of skeletal muscle have many nuclei and striations. Skeletal muscle is the only type of muscle under *voluntary* control through the nervous system (see Chapter 16), which means you can decide when you want to contract a skeletal muscle.

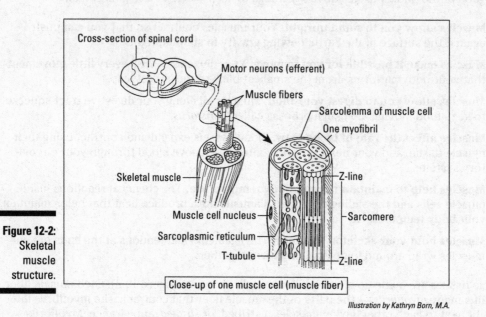

Cross-section of spinal cord

Motor neurons (efferent)

Muscle fibers

Sarcolemma of muscle cell

One myofibril

Skeletal muscle

Z-line

Muscle cell nucleus

Sarcomere

Sarcoplasmic reticulum

T-tubule

Z-line

Close-up of one muscle cell (muscle fiber)

Figure 12-2:
Skeletal
muscle
structure.

Illustration by Kathryn Born, M.A.

38. Put the following terms in order from largest to smallest. Then circle the word that represents one muscle cell.

a. Myofibril

b. Skeletal muscle tissue

c. Muscle fiber

d. Actin (thin) filament

e. Sarcomere

39.–43. Use the terms that follow to answer the following questions and identify which muscle type has each characteristic.

a. Cardiac muscle

b. Smooth muscle

c. Skeletal muscle

39. This muscle type is under voluntary control.

40. This muscle type has cells with multiple nuclei per cell (fiber).

41. This muscle type controls digestion.

42. This muscle type is only found in one organ in the body.

43. This muscle type lacks obvious striations.

Answers to Questions on the Skeletal and Muscular Systems

The following are answers to the practice questions presented in this chapter.

1 The answer is **a. Gravity.**

2 The answer is **a. Gravity.**

3 The answer is **b. Friction (resistance).**

4 The answer is **b. Friction (resistance).**

5 The answer is **a. Gravity.**

6 The answer is **c. Endoskeleton.**

7 The answer is **b. Exoskeleton.**

8 The answer is **a. Hydrostatic skeleton.**

9 The answer is **c. Endoskeleton (but made of cartilage, not bone).**

10 The answer is **c. Endoskeleton.**

11 Hydrostatic skeleton: Advantage is it's very flexible; disadvantage is it's not as good at resisting gravity as harder skeletons, and it doesn't protect soft parts. Exoskeleton: Advantage is that it gives good protection to an animal's soft parts; disadvantage is that it's restrictive to growth. Animals with exoskeletons have to shed their skeletons in order to grow larger. Endoskeleton: Advantage is that it gives good support against gravity; disadvantage is that it doesn't protect soft parts as well as exoskeletons.

12 You should have shaded the skull, ribs, and spinal column in one color (axial) and the rest of the bones in another color (appendicular).

13 – 30 The following is how Figure 12-1 should be labeled:

13 **b. Skull**; 14 **e. Vertebrae**; 15 **q. Pectoral girdle**; 16 **o. Ribs**; 17 **f. Sternum**; 18 **a. Humerus**; 19 **i. Radius**; 20 **j. Ulna**; 21 **r. Carpals**; 22 **c. Metacarpals**; 23 **g. Phalanges of the hand**; 24 **d. Pelvic girdle**; 25 **k. Femur**; 26 **m. Tibia**; 27 **h. Fibula**; 28 **l. Tarsals**; 29 **p. Metatarsals**; 30 **n. Phalanges of the foot.**

31 The answer is **a. Bone matrix.**

32 The answer is **d. Yellow bone marrow.**

33 The answer is **c. Red bone marrow.**

34 The answer is **e. Connective tissue.**

35 The answer is **b. Pivot joint.**

You can turn your head and angle it, but you can't spin it completely around.

36 The answer is **c. Hinge joint.**

You can fold your lower arm upward onto your upper arm, just like opening and closing a book.

37 The answer is **a. Ball-and-socket joint.**

You can rotate your leg in all directions within your hip socket.

38 The answer is **b. Skeletal muscle tissue; c. Muscle fiber; a. Myofibril; e. Sarcomere; d. Actin (thin) filament.** You should have circled **c. Muscle fiber** because that's the name for a muscle cell.

39 The answer is **c. Skeletal muscle.**

40 The answer is **c. Skeletal muscle.**

41 The answer is **b. Smooth muscle.**

42 The answer is **a. Cardiac muscle** (found only in the heart!).

43 The answer is **b. Smooth muscle.**

Chapter 13

Giving Your Body What It Needs: The Respiratory and Circulatory Systems

All living things need to be able to exchange materials — like food, oxygen, and waste products — with the environment and then circulate those materials around their bodies. In animals, respiratory systems enable the uptake and exchange of gases, and circulatory systems move nutrients and gases around the body. In this chapter, I present an introduction to the diversity of animal systems and then focus on the details of the human respiratory and circulatory systems.

Catching Your Breath: Animal Respiration

Respiration is the exchange of life-sustaining gases, such as oxygen, between an animal and its environment. Gas exchange occurs by diffusion (see Chapter 3 for details), moving necessary gases like oxygen into animals and taking away waste gases like carbon dioxide. Although animals have different ways of moving gases in and out of their bodies, gas exchange between the animal and its environment occurs across a moist surface.

Most animal respiration involves four steps:

1. **Taking air in *(breathing* or *inspiration).***

2. **Circulating gases throughout the body.**

3. **Exchanging needed gases for unnecessary gases.**

4. **Using the needed gases (in cellular respiration; see Chapter 4).**

Depending on the complexity of their bodies and the environment in which they live, animals evolved different systems to achieve respiration. Four basic types of gas-exchange systems occur in animals:

✔ **Integumentary exchange** occurs through the outer surface of some small animals that constantly stay moist.

✔ **Gills** are structures that extend outward from an animal's body to exchange gases in watery environments.

✔ **Tracheal exchange systems** rely on a network of tubes that end in holes to move oxygen and carbon dioxide throughout the bodies of certain types of insects.

✔ **Lungs** are structures that extend into an animal's body, creating moist internal surfaces that use diffusion to transport gases into and out of the body.

1.–4. Use the terms below to identify which part of respiration is blocked by the condition stated in the question.

   a. Breathing

   b. Circulation

   c. Gas exchange

   d. Cellular respiration

1. A girl with anemia doesn't have enough of the protein hemoglobin to carry oxygen in her red blood cells.

2. A boy with exercise-induced asthma suffers narrowing of his airways during exercise.

3. A woman has emphysema, which damages the air sacs of her lungs, reducing the amount of surface area for gas exchange.

4. A man accidentally swallows a large piece of food that extends his food tube so that it pinches off his windpipe.

5.–8. Use the following terms to label each type of gas-exchange system in Figure 13-1. Also, state one example of an animal that would have each system.

   a. Integumentary exchange

   b. Lungs

   c. Gills

   d. Tracheal exchange system

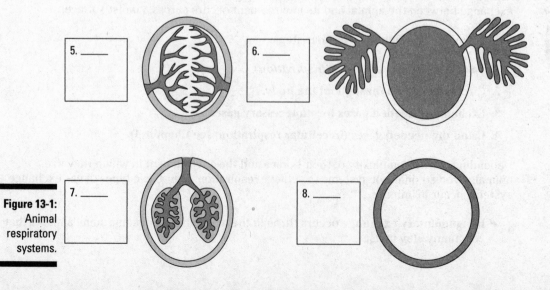

Figure 13-1: Animal respiratory systems.

5. _____

6. _____

7. _____

8. _____

Taking a Breath with the Human Respiratory System

Humans have a pair of lungs that lie in the chest cavity. When you breathe in *(inspiration or inhalation),* your diaphragm muscle, which sits below your lungs, contracts, becoming smaller and moving downward, which allows your rib cage to move upward and outward. Because the lungs have more room when your chest is expanded, they open up, and air rushes in to fill the space. When your diaphragm relaxes, your rib cage moves back downward and inward, increasing air pressure inside your lungs and forcing air out *(expiration* or *exhalation).* As you breathe in, air moves through the chambers of your respiratory system in the following order:

1. **Air enters through your nostrils and into your *nasal cavity* and then flows into the top part of your throat.**

   The nasal cavity warms, moistens, and filters (through hair and mucus) the air as it passes through.

2. **Air then moves into the middle part of your throat, called your *pharynx*.**

   You can recognize your pharynx because it starts out larger near your nasal cavity and tapers into a narrow space where it connects to your food tube (see Chapter 14 for more on your digestive system).

3. **Air then flows through the space around your vocal cords, which is called your *larynx*.**

   In men, the cartilage around the larynx sticks out so much it forms the bulge in men's necks called the *Adam's apple*.

4. **Air enters your windpipe, or *trachea*.**

   You can recognize your trachea by its C-shaped rings of strong cartilage and connective tissue that provide strength and help keep your trachea open.

5. **The trachea splits into two *bronchi* just above the heart.**

   Air moves through your bronchi and into a network of increasingly branched and smaller passages until it reaches fine tubes called *bronchioles*.

6. **The air reaches the end of its journey in little clusters of sacs called *alveoli* that look a little bit like raspberries.**

   Each *alveolus* is wrapped with capillaries so that gas exchange can occur between the lungs and the blood.

9.–15. Use the following terms to label the structures of the human respiratory system in Figure 13-2.

   a. Bronchiole

   b. Alveolus

   c. Nasal cavity

   d. Larynx

   e. Bronchus

   f. Pharynx

   g. Trachea

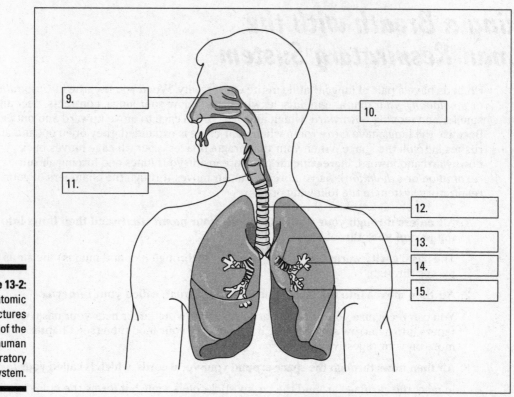

Figure 13-2:
Anatomic
structures
of the
human
respiratory
system.

9.

10.

11.

12.

13.

14.

15.

From LifeART®, Super Anatomy 1, © 2002, Lippincott Williams & Wilkins

16. Imagine that you're removing carbon dioxide from your body. Beginning with the delivery of carbon dioxide to the lungs, trace the pathway of carbon dioxide by putting the terms in the following list in the order in which carbon dioxide would pass through on its way out of your body: larynx, bronchus, trachea, pharynx, bronchiole, alveolus, nasal cavity.

In with the Good, Out with the Bad: Animal Circulatory Systems

Every animal alive possesses a circulatory system that's in charge of bringing nutrients to cells and removing wastes so they don't cause disease. While they move fluids around the body, circulatory systems also help out with other tasks by

✔ Delivering oxygen to cells and picking up carbon dioxide

✔ Distributing hormones to cells

✔ Maintaining body temperature by transporting heat

✔ Transporting cells to fight infection (more on this in Chapter 15)

Animals have two types of circulatory systems:

✔ In **open circulatory systems,** the animal's heart pumps a bloodlike fluid called *hemolymph* through open-ended vessels into a chamber called the *hemocoel,* where it directly bathes the cells. In other words, the circulatory system's fluid isn't kept separate from the fluid around cells, which is called *interstitial fluid.* Muscle contractions push the hemolymph back toward the heart so it can be circulated throughout the animal again and again.

✔ In **closed circulatory systems,** a network of closed tubes called *vessels* performs the transportation and prevents blood from coming into direct contact with the body's cells. Three types of vessels move blood within closed systems:

- *Arteries* carry blood from the heart to the organs and tissues of the body.

- *Veins* carry blood from the organs and tissues back to the heart.

- *Capillaries* are fine networks of vessels that connect arteries and veins within each tissue.

Animal *hearts* come in different sizes and shapes, but they have the same function: to pump fluid throughout the circulatory system. That fluid is either hemolymph or blood, depending on the type of circulatory system.

17.–18. Use the following terms to identify the type of circulatory system described in each situation.

a. Open circulatory system

b. Closed circulatory system

17. In the grasshopper, the heart pumps fluid through an aorta that runs along the insect's dorsal side. From there, fluid moves into chambers called *sinuses* where it comes into contact with body cells. Contractions of body muscles push the fluid back toward the heart.

18. An octopus has three hearts. Two hearts pump fluid through vessels to the gills so that the fluid can receive oxygen and release wastes, and then they pump the blood through vessels to the third heart. The third heart pumps the oxygen-rich fluid through vessels to the rest of the body.

Navigating the Human Heart and Circulatory System

The heart and circulatory system of a human, as well as some other mammals, are complex. These large animals need to have a higher blood pressure to push the blood throughout their entire bodies. This need results in a *two-circuit circulatory system,* a system that has two distinct pathways:

✔ One pathway is for **pulmonary circulation,** which first delivers deoxygenated blood to the lungs so it can become oxygenated and then delivers oxygenated blood back to the heart.

✔ The other circuit is for **systemic circulation,** which carries oxygenated blood from the heart to the rest of the body and back.

The human heart has four chambers:

- ✔ Two **ventricles,** muscular chambers that squeeze blood out of the heart and into the blood vessels. They reside at the bottom of the heart.

- ✔ Two **atria,** muscular chambers that drain and then squeeze blood into the ventricles. They reside at the top of the heart.

Your heart is divided into halves because of your two-circuit circulatory system, so you have a left atrium and a left ventricle, as well as a right atrium and a right ventricle. Your heart's right side pumps blood to your lungs, and its left side pumps blood to the rest of your body. Blood goes into both pathways with each and every pump. *Valves* separate one chamber of the heart from another. Each valve consists of strong flaps of muscle tissue, called *cusps* or *leaflets.* When your heart is working properly, the valves open and close fully so blood can only flow in one direction through it.

Four valves separate the four chambers of your heart from one another and from the major blood vessels that are connected to it (see Figure 13-3):

- ✔ The **right atrioventricular (AV) valve** is located between the right atrium and the right ventricle. This valve is also referred to as the **tricuspid valve** because it has three flaps in its structure.

- ✔ The **pulmonary semilunar valve** separates the right ventricle from the pulmonary artery. *Semilunar* means "half-moon" and refers to the valve's shape.

- ✔ The **left atrioventricular (AV) valve** is located between the left atrium and the left ventricle. This valve is also called the **bicuspid valve** because it has only two flaps in its structure.

- ✔ The **aortic semilunar valve** separates the left ventricle from the aorta. Like the pulmonary semilunar valve, this valve has a half-moon shape.

19.–26. Use the terms that follow to label the valves and chambers of the human heart in Figure 13-3.

   a. Left AV valve

   b. Right ventricle

   c. Pulmonary semilunar valve

   d. Right atrium

   e. Left atrium

   f. Aortic semilunar valve

   g. Left ventricle

   h. Right AV valve

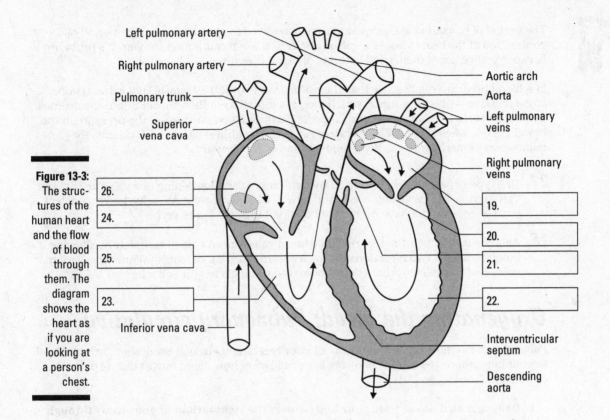

Figure 13-3: The structures of the human heart and the flow of blood through them. The diagram shows the heart as if you are looking at a person's chest.

Left pulmonary artery

Right pulmonary artery

Pulmonary trunk

Superior vena cava

26.

24.

25.

23.

Inferior vena cava

Aortic arch

Aorta

Left pulmonary veins

Right pulmonary veins

19.

20.

21.

22.

Interventricular septum

Descending aorta

Entering the cardiac cycle

Your heart is an impressive little organ. Even though it's only as big as a clenched adult fist, it pumps 5 liters of blood throughout your body 70 times a minute. Your heart never stops working from the time it starts beating in the embryo until the moment you die. It doesn't even get an entire second to rest. It beats continually every 0.8 seconds of your life.

REMEMBER

The eight-tenths of a second that a heart beats is called the *cardiac cycle.* During that 0.8-second period, your heart forces blood into your blood vessels and then takes a quick rest (for just 0.4 seconds). Here's what happens in those 0.8 seconds:

✔ **Contraction of the left and right atria:** This contraction squeezes blood down into the ventricles.

✔ **Contraction of the left and right ventricles:** This contraction forces blood into the blood vessels that leave the heart.

✔ **Resting of the atria and ventricles:** The relaxed atria allow the blood within them to drain into the ventricles.

The period of relaxation in the heart muscle is referred to as *diastole,* and the period of contraction in the heart muscle is called *systole.* If these terms sound familiar, it's probably because you've heard them used in terms of blood pressure.

In a blood pressure reading, such as the normal value of 120/80 mmHg, 120 is the *systolic blood pressure* — the pressure at which blood is forced from the ventricles into the arteries when the ventricles contract — and 80 is the *diastolic blood pressure* — the pressure in the blood vessels when the muscle fibers are relaxed. The abbreviation *mmHg* stands for millimeters of mercury (Hg is the chemical symbol for mercury).

27. One type of heart defect is caused by a thickening and hardening of a valve so that the valve has greater resistance to blood flow. If a person's right AV valve had this defect, what effect would it have on the flow of blood through her heart?

28. Another type of heart defect results when a valve doesn't close completely, allowing blood to flow in two directions during a contraction. If a person's pulmonary semilunar valve had this defect, what effect would it have on his health and why?

Oxygenating the blood: Pulmonary circulation

Pulmonary circulation, the first pathway of your two-circuit circulatory system, brings blood to your lungs for oxygenation. Following is a rundown of how blood moves during pulmonary circulation (see Figure 13-4):

1. **Deoxygenated blood from your body enters the right atrium of your heart through the *superior vena cava* and the *inferior vena cava.***

   Superior means "higher," and *inferior* means "lower," so the superior vena cava is at the top of the right atrium, and the inferior vena cava is at the bottom of the right atrium.

2. **From the right atrium, the deoxygenated blood drains into the right ventricle through the right AV valve.**

   When the ventricles contract, the right AV valve closes off the opening between the ventricle and the atrium so blood doesn't flow back into the atrium.

3. **The right ventricle then contracts, forcing the deoxygenated blood through the pulmonary semilunar valve and into the pulmonary artery.**

   The pulmonary semilunar valve keeps blood from flowing back into the right ventricle after it's in the pulmonary artery.

4. **The pulmonary artery carries the blood that's very low in oxygen to the lungs, where it becomes oxygenated.**

5. **Freshly oxygenated blood returns from the lungs to the heart via the pulmonary veins.**

   Note that your pulmonary veins are the only veins in your body that contain oxygenated blood; all your other veins contain deoxygenated blood.

29. Use a colored pencil or highlighter to shade the pulmonary pathway of your circulatory system in Figure 13-4.

30.–33. Use the following terms to label the structures of the pulmonary pathway of your circulatory system in Figure 13-4.

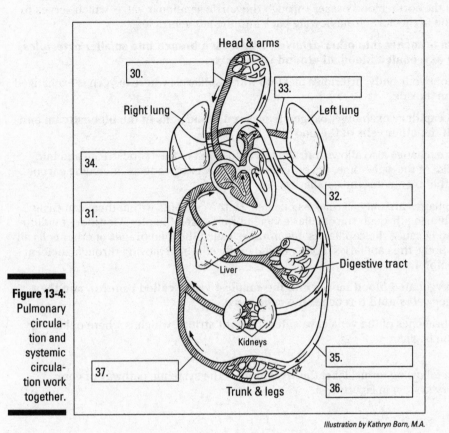

Head & arms

30.

33.

Right lung

Left lung

34.

32.

31.

Liver

Digestive tract

Figure 13-4:
Pulmonary
circula-
tion and
systemic
circula-
tion work
together.

Kidneys

35.

37.

36.

Trunk & legs

Illustration by Kathryn Born, M.A.

a. Pulmonary artery

b. Pulmonary vein

c. Superior vena cava

d. Inferior vena cava

Spreading oxygenated blood around: Systemic circulation

Systemic circulation brings oxygenated blood to your body's cells. Here's how blood moves through this pathway (refer to Figure 13-4):

1. **The pulmonary veins push blood into the left atrium.**

   When the left atrium relaxes, the oxygenated blood drains into the left ventricle through the left AV valve.

2. **As the left ventricle contracts, the oxygenated blood is pumped into the main artery of the body — the aorta.**

   To get to the aorta, blood passes through the aortic semilunar valve, which serves to keep blood in the aorta from flowing back into the left ventricle.

3. **The aorta branches into other *arteries,* which then branch into smaller *arterioles,* carrying oxygenated blood all around your body.**

   Throughout your body, arterioles meet up with capillaries where oxygen is exchanged for carbon dioxide.

4. **Through capillary exchange, oxygen leaves red blood cells in the bloodstream and enters all the other cells of the body.**

   Capillary exchange also allows nutrients to diffuse out of the bloodstream and into other cells. At the same time, the other cells expel waste products, including carbon dioxide, that then enter the capillaries.

   Your capillaries are only as thick as one cell, so the contents within them can easily exit by diffusing through the capillaries' membranes (see Chapter 3 for more on diffusion). And because the capillaries' membranes touch the membranes of other cells all over the body, the capillaries' contents can easily continue moving through adjacent cells' membranes.

5. **The deoxygenated blood moves into the smallest veins, called *venules,* and then into bigger *veins* until it reaches the *vena cava.***

   The two branches of the vena cava enter the right atrium, which is where pulmonary circulation begins.

34.–37. Using the following terms, label the structures of the systemic pathway of your circulatory system in Figure 13-4.

   e. Aorta

   f. Vein

   g. Artery

   h. Capillaries

38. Using a different color ink than you used to mark your pulmonary circulatory pathway, shade in your systemic circulatory pathway in Figure 13-4.

39. Place the following terms in order, ranking them from vessels that carry blood with the least oxygen to those that carry blood with the most oxygen.

   a. Venules

   b. Capillaries

   c. Arteries

   d. Veins

   e. Arterioles

40. Imagine that you're a blood cell in the kidneys. In sequential order, name all the chambers and valves of the heart you'll travel through between the time you leave the kidney and the time you return.

Answers to Questions on the Respiratory and Circulatory Systems

The following are answers to the practice questions presented in this chapter.

1 The answer is **b. Circulation.**

Circulation is affected because she can't effectively circulate the oxygen she breathes.

2 The answer is **a. Breathing.**

Breathing is affected because he can't draw in enough air.

3 The answer is **c. Gas exchange.**

Gas exchange is affected because she doesn't have enough surface area in her lungs.

4 The answer is **a. Breathing.**

Breathing is affected because he can inhale any air.

5 The answer is **d. Tracheal exchange system.**

The tracheal exchange systems occur in insects like grasshoppers.

6 The answer is **c. Gills.**

Fish and lobsters are two examples of organisms with gills.

7 The answer is **b. Lungs.**

Lungs occur in animals like mammals and birds.

8 The answer is **a. Integumentary exchange.**

Earthworms use integumentary exchange.

9–15 The following is how Figure 13-2 should be labeled:

9 c. **Nasal cavity;** 10 f. **Pharynx;** 11 d. **Larynx;** 12 g. **Trachea;** 13 e. **Bronchus;** 14 a. **Bronchiole;** 15 b. **Alveolus.**

16 The answer is **Alveolus→Bronchiole→Bronchus→Trachea→Larynx→Pharynx→Nasal cavity.**

17 The answer is **a. Open circulatory system.**

The fluid flows into chambers and has direct contact with body cells.

18 The answer is **b. Closed circulatory system.**

The fluid is contained in vessels as it travels around the body.

19–26 The following is how Figure 13-3 should be labeled:

19 e. **Left atrium;** 20 a. **Left AV valve;** 21 f. **Aortic semilunar valve;** 22 g. **Left ventricle;** 23 b. **Right ventricle;** 24 h. **Right AV valve;** 25 c. **Pulmonary semilunar valve;** 26 d. **Right atrium.**

27 Blood flow would be restricted and would accumulate in the right atrium. This would increase pressure on the heart during contractions.

28 During the contraction of the heart, blood wouldn't flow effectively out of the right ventricle and into the pulmonary artery. This would result in less effective oxygenation of the blood, leading to symptoms like shortness of breath and tiredness (because the person wouldn't be getting enough oxygen).

29 You should have shaded the upper loop in Figure 13-4 from the heart to the lungs and back to the heart.

30–**33** The following is how Figure 13-4 should be labeled:

30 c. **Superior vena cava;** 31 d. **Inferior vena cava;** 32 a. **Pulmonary artery;** 33 b. **Pulmonary vein.**

34–**37** The following is how Figure 13-4 should be labeled:

34 e. **Aorta;** 35 g. **Artery;** 36 h. **Capillaries;** 37 f. **Vein.**

38 You should have shaded the entire rest of the circulatory system (except for the loop from the heart to the lungs and back to the heart).

39 The answer is **Veins→Venules→Capillaries→Arterioles→Arteries.**

40 The answer is **Kidneys→Capillary→Venule→Vein→Inferior Vena Cava→Right atrium→ Right AV valve→Right ventricle→Pulmonary semilunar valve→Pulmonary artery→ Capillaries in lungs→Pulmonary vein→Left atrium→Left AV valve→Left ventricle→ Aortic semilunar valve→Aorta→Artery→Arteriole→Capillaries in kidneys.**

Chapter 14

Processing Food with the Digestive and Excretory Systems

....

In This Chapter

▶ Explaining the parts and functions of the human digestive system

▶ Breaking down the urinary system

▶ Understanding the kidney's structure and functions

....

After an animal ingests or absorbs food, its digestive system immediately starts breaking down the food to release the nutrients within it. After the animal absorbs the useful nutrients into the bloodstream, the animal eliminates solid wastes through its large intestine and liquid wastes through its urinary system. In this chapter, I present the workings of animal digestive systems. In addition, I take you on a tour of your own digestive and excretory systems and give you an opportunity to practice thinking about their structure and function.

Got Food? Understanding What Happens When Animals Eat

All animals need to break down food molecules into smaller pieces so they can circulate them around their bodies to all their cells. Their cells take in small food molecules and use them as material for growth or as a source of energy (see Chapter 4).

Four main events occur from the moment food enters an animal's body until the time the animal releases its wastes:

✔ **Ingestion** occurs when an animal takes food into its digestive tract.

✔ **Digestion** occurs when the animal's body gets busy breaking down the food. Two types of digestion exist in all animals:

- **Mechanical digestion** physically breaks down food into smaller and smaller pieces. It begins when an animal consumes the food and continues until the food enters the animal's stomach.

- **Chemical digestion** uses enzymes and acids to break down chewed or ground-up food into even smaller pieces. It also begins as soon as an animal consumes the food as the mouth's enzymes go to work. Chemical digestion continues as the food moves through the stomach and small intestine and encounters enzymes and acids in the stomach and enzymes in the small intestine.

✔ **Absorption** occurs when cells within the animal move small food molecules from the digestive system to the insides of the cells.

✔ **Elimination** occurs when material that the animal can't digest passes out of its digestive tract.

The basic way an animal's digestive system works has a great deal to do with whether it can spend a few hours between meals or whether it has to keep consuming food constantly just to stay alive. Animals with *incomplete digestive tracts* have the most primitive digestive systems. These animals have a gut with just one opening that serves as both mouth and anus. Animals like humans have more complex systems that scientists call a *complete digestive tract*. Complete digestive tracts have a mouth at one end and an anus at the other.

Complete digestive tracts are more efficient than incomplete digestive tracts because they allow thorough digestion of food before excretion occurs. Organisms with incomplete digestive tracts release undigested food along with their wastes, so they often have to take in food constantly to replace food that's excreted before they extract all the nutrients:

✔ Animals that must consume constantly because they take food in and then push it out soon afterward are called *continuous feeders*. Most of these animals are either permanently attached to something (think clams or mussels) or incredibly slow movers.

✔ Animals that are *discontinuous feeders* consume larger meals and store the ingested food for later digestion. These animals are generally more active and somewhat nomadic.

You and all the other animals that are discontinuous feeders must have a place in the body to store food as it slowly digests. In humans, this organ is the *stomach*.

1.–6. Use the following terms to identify which step of digestion is occurring in each example.

    a. Ingestion

    b. Digestion

    c. Absorption

    d. Elimination

1. The enzyme salivary amylase breaks down starch molecules in the mouth into smaller simple sugars.

2. A sea gull releases feces as it flies over a dock.

3. A ball python swallows a mouse.

4. A giraffe pulls some leaves from a tall tree with its tongue.

5. The cells that line your small intestine transport amino acids from the solution in your intestines into their cytoplasm.

6. Your pancreas releases *lipases*, enzymes that break down fat, into your small intestine where they break down lipids into smaller molecules.

7.–9. Use the following terms to identify the type of digestive system described in each example.

    a. Incomplete digestive system

    b. Complete digestive system

7. An anemone catches some food particles with its sticky tentacles and places them in its mouth. The food travels down into the anemone's body cavity, where digestion occurs. The anemone releases indigestible bits of food from its mouth.

8. A hamster nibbles on sunflower seeds. After swallowing and digesting the seeds, the hamster releases fecal pellets from its anus.

9. A grasshopper chews some wheat with its mouth parts and then swallows the wheat. The grasshopper digests the wheat in its stomach and then absorbs small food molecules using the cells that line the stomach. In the hind gut, the grasshopper reabsorbs water from the remains of the food, concentrating the indigestible parts into fecal pellets that it releases from its anus.

Moving Along the Human Digestive System

Humans have a complete digestive tract: Food enters at one end and wastes exit from the opposite end. Digestion begins in the mouth and continues as food moves through your system (see Figure 14-1 later in this section to follow along with an illustration):

✔ Digestion in the mouth occurs by both chemical and mechanical means.

- Chewing, or *mastication,* mechanically breaks food into smaller pieces.

- Your taste buds stimulate the production of saliva to help moisten the food, physically preparing it for you to swallow it.

- Saliva contains the enzyme *salivary amylase,* which chemically digests the complex carbohydrate starch into simple sugars (glucose).

✔ Your tongue pushes the chewed food to the back of your throat toward your pharynx, the muscular chamber at the back of your throat. As you swallow, your palate raises until it's pressed up against the wall of your pharynx, preventing food from entering your nasal cavity (unless someone makes you laugh while you're swallowing!).

✔ Your muscles squeeze, in a process called *peristalsis,* the food mass, or *bolus,* into your *esophagus,* the tube that connects your mouth to your stomach.

✔ In the stomach, peristalsis continues and *gastric juices* chemically digest the food to a thick liquid called *chyme.* Gastric juice is extremely acidic, with a pH range between 1 and 4, and contains the enzyme *pepsin,* which breaks proteins into smaller chains of amino acids (see Chapter 2 for more on proteins).

✔ Chyme passes through the *pyloric valve,* the gate between your stomach and small intestine, and into your small intestine. Your *pyloric sphincter* muscle occasionally opens the valve, allowing your stomach's contents into your small intestine a little bit at a time. Food arrives at your small intestine between one and four hours after you eat. After food molecules hit your small intestine, your liver and pancreas break them down into even smaller units:

- Your *liver* is the largest gland in your body. It's a large, lobed structure that wraps around the *gallbladder,* a small, pear-shaped structure. The liver secretes diluted *bile* into the gallbladder, which stores and concentrates the bile, and then releases it into the small intestine. Bile, a liquid mixture secreted by the liver, helps to *emulsify* fats so they're suspended in water and you can digest them more easily. (If you've ever shaken up a bottle of salad dressing and forced oil to break up into small droplets that mix with the water portion of the dressing, you have first-hand experience with emulsification.)

- Your *pancreas* has an irregular, almost triangular shape that begins with a larger end near the junction between the stomach and small intestine. Your pancreas releases *pancreatic juice* into your small intestines, contributing a mix of digestive enzymes to help chemically digest food molecules: *Lipase* breaks apart fat molecules, *pancreatic amylase* breaks apart long carbohydrates, and the enzymes *trypsin* and *chymotrypsin* break apart peptide fragments.

Don't let the word *small* fool you. The small intestine is much longer than the large intestine (over 20 feet long versus about 5 feet long). The term *small intestine* refers to the fact that this part of the intestines is narrower in diameter than the large intestine; the large intestine is wider in diameter but shorter in length.

- ✔ **Your small intestine is the primary site of absorption of small food molecules into your cells.** Your body absorbs the nutrients it can use into the cytoplasm of the cells lining your small intestine.

- ✔ **The rest of the material that you can't further digest or use passes on to the *large intestine,* or *colon.*** The large intestine absorbs water back into your body, concentrating the waste material into *feces*. Feces pass through your *rectum* and leave your body through the *anus*. A small, worm-like appendage called the *appendix* dangles off one part of your colon. For a long time, scientists thought the appendix had no function, but recent research suggests that it plays a role in immunity.

10.–20. Use the terms that follow to identify the parts of the human digestive system shown in Figure 14-1.

   a. Small intestine

   b. Anus

   c. Stomach

   d. Salivary glands

   e. Liver

   f. Pancreas

   g. Rectum

   h. Appendix

   i. Gallbladder

   j. Large intestine (colon)

   k. Esophagus

21.–27. Use the following terms to match the parts of the digestive system to their role in the process of digestion. Some parts may have more than one function.

   a. Ingestion

   b. Digestion

   c. Absorption

   d. Elimination

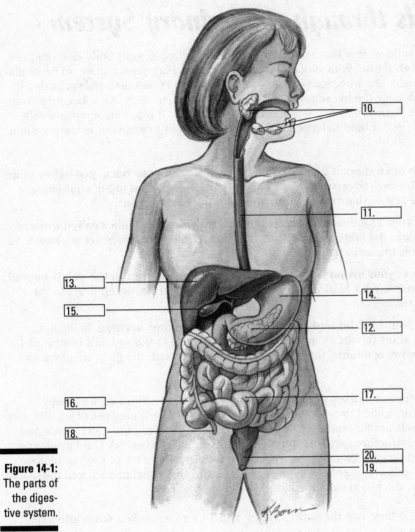

10.

11.

13.

15.

14.

12.

16.

17.

18.

20.

19.

Figure 14-1:
The parts of
the diges-
tive system.

Illustration by Kathryn Born, M.A.

21. Salivary glands

22. Stomach

23. Small intestine

24. Large intestine (colon)

25. Liver

26. Pancreas

27. Anus

Filtering Fluids through the Urinary System

In addition to the solid wastes that your body eliminates as feces, your body also releases wastes that are filtered from your blood and tissue fluids as part of your *urine*. In particular, urine helps you flush out *nitrogenous wastes* — unnecessary, excess materials containing nitrogen that result from the breakdown of proteins and nucleic acids. Also, because your *urinary system* (see Figure 14-2) releases fluid from your body, it plays an important role in maintaining the proper fluid balance in the body. The structures of your urinary system work like this:

- ✔ **You have two bean-shaped *kidneys,* one on each side of your back, just below your ribs.** Your kidneys produce urine. The *adrenal glands* that sit on top of your kidneys are endocrine glands that release hormones into the bloodstream.

- ✔ **Urine leaves your kidneys and travels through thin muscular tubes called *ureters.*** Smooth muscle in the ureter pushes the urine along through peristalsis (see Chapter 12 for more details on smooth muscle).

- ✔ **Urine arrives at your urinary *bladder,* a muscular bag that lies in the pelvis behind your pubic bones.** Your bladder stretches as it fills with urine, which leads to the signal to urinate.

- ✔ **Urine leaves your bladder and exits the body through your *urethra.*** In females, the urethra is short (about 1½ inches long) and lies close to the vagina's front wall. In males, the urethra is about 8 inches long and passes through the prostate gland and penis.

Your kidneys are the workhorses of your urinary system. Blood arrives at the kidney through a large artery called the *renal artery* and then passes into a network of smaller and smaller blood vessels until it reaches a system of capillaries that are intimately entwined within the kidney's structure (see Chapter 13 for more on blood vessels). Fluid from your capillaries is forced by your blood pressure into the kidney, where it's filtered to remove wastes. After the fluids are filtered, the kidney returns the clean portions back to the circulatory system, and blood exits the kidneys via the *renal vein.*

The function of the kidney, and the entire urinary system, can be broken down into four important components:

- ✔ Filtration of body fluids by passing them through the kidney.

- ✔ Reabsorption of useful materials like water and *electrolytes* (charged ions such as Na$^+$ and K$^+$) from the kidney back to the blood.

- ✔ Secretion of specific waste materials from the blood into the kidney.

- ✔ Excretion of wastes in urine.

28.–31. Use the terms that follow to identify the parts of the human urinary system shown in Figure 14-2.

    a. Bladder

    b. Kidney

    c. Ureter

    d. Urethra

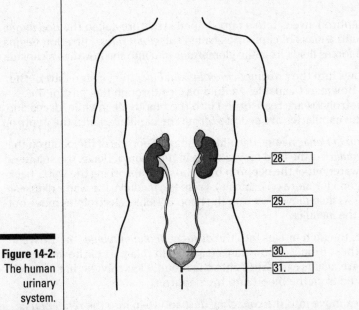

From LifeART®, Super Anatomy 1, © 2002, Lippincott Williams & Wilkins

Figure 14-2: The human urinary system.

28.

29.

30.

31.

Exploring the Inner Workings of the Human Kidney

Like most organs in the human body, the function of a kidney is closely tied to its structure. The outer covering on each kidney, called the *capsule,* is made of stretchy collagen fibers that help anchor your kidneys. Under the capsule, each kidney has three distinct areas:

✔ The **renal cortex,** which is the outer layer.

✔ The **renal medulla,** which is the middle layer. The renal medulla folds into cone-shaped projections called *renal pyramids.*

✔ The **renal pelvis,** the inner layer that tapers and becomes a ureter.

Each kidney contains more than 1 million *nephrons,* microscopic tubules that make urine. Each nephron contributes to a collecting duct that carries the urine into the renal pelvis and then down the ureter.

Each of the million tiny nephrons in one of your kidneys is a mass of tiny, looped tubules that begin and end in the renal cortex. The nephrons are closely associated with capillaries in the kidney. Fluid from the capillaries enters the nephrons at their *proximal* (near) end. Wastes are filtered from the fluid as it passes through the nephron, and then the useful water and electrolytes are returned to the blood. Concentrated wastes leave the nephrons at their *distal* (far) end and then enter a collecting duct that empties into the renal pelvis.

The structure of a nephron is closely tied to its function as a filter:

- ✔ At the distal end, the nephron swells into a cup-shaped structure called the *Bowman's capsule* that wraps around a mass of capillaries called the *glomerulus*. Filtration begins as pressure in the blood forces fluids from the glomerulus and into the Bowman's capsule.

- ✔ Fluid in the nephron flows into the *proximal convoluted tubule,* the twisted part of the nephron closest to the Bowman's capsule. As fluid passes through this part of the nephron, water and electrolytes are reabsorbed into the blood. Meanwhile, drugs and toxins that are still in the capillaries are secreted from the capillaries into the nephron.

- ✔ Fluid passes into the *Loop of Henle,* a long, dangling, U-shaped portion of the nephron that passes into the *renal medulla*. As the fluid moves down in the Loop of Henle, the solutes in the renal medulla draw water out of the nephron by osmosis, reabsorbing the water back into the body (see Chapter 3 for more on osmosis). From the medulla, the water diffuses back into the capillaries. As fluid moves back up the Loop of Henle, electrolytes move out of the nephron and into the medulla.

- ✔ From the Loop of Henle, the fluid moves into the *distal convoluted tubule*, the twisted part of the nephron farthest from the Bowman's capsule. In this part of the nephron, water and electrolytes are again reabsorbed into the blood. Electrolytes that maintain blood pH may be secreted from the blood into the nephron.

- ✔ The concentrated wastes move into the *collecting duct* and then into the *renal pelvis.* More water is reabsorbed as the fluid passes through the collecting duct. From the renal pelvis, the wastes move into the ureter and down to the bladder.

Urine is continuously spurted from the ureter into the top of the bladder. Although the bladder can hold up to a pint of urine, you typically begin to feel the need to urinate when your bladder is only one-third full. When your bladder is two-thirds full, you start to feel really uncomfortable. When you want to start urinating, the sphincter muscle at the top of your urethra relaxes, opening the urethra and letting the urine out.

Your kidneys do an amazing job of concentrating your wastes. For every 125 milliliters of fluid that leaves your blood every minute, only one milliliter of fluid leaves your kidneys to enter your bladder. The rest of the fluid is recycled back to your blood!

32.–36. Use the terms that follow to match each structure with the function it performs. Some structures may perform more than one function.

   a. Filtration

   b. Reabsorption

   c. Secretion

   d. Excretion

32. Bowman's capsule

33. Proximal convoluted tubule

34. Loop of Henle

35. Distal convoluted tubule

36. Collecting duct

Answers to Questions on the Digestive and Excretory Systems

The following are answers to the practice questions presented in this chapter.

1 The answer is **b. Digestion.**

2 The answer is **d. Elimination.**

3 The answer is **a. Ingestion.**

4 The answer is **a. Ingestion.**

5 The answer is **c. Absorption.**

6 The answer is **b. Digestion.**

7 The answer is **a. Incomplete digestive system.**

8 The answer is **b. Complete digestive system.**

9 The answer is **b. Complete digestive system.**

10–20 The following is how Figure 14-1 should be labeled:

10 **d. Salivary glands;** 11 **k. Esophagus;** 12 **f. Pancreas;** 13 **e. Liver;** 14 **c. Stomach;**
15 **i. Gallbladder;** 16 **j. Large intestine (colon);** 17 **a. Small intestine;** 18 **h. Appendix;**
19 **b. Anus;** 20 **g. Rectum.**

21 The answer is **b. Digestion.**

22 The answer is **b. Digestion.**

23 The answer is **b. Digestion** and **c. Absorption.**

24 The answer is **d. Elimination** (absorption of water).

25 The answer is **b. Digestion.**

26 The answer is **b. Digestion.**

27 The answer is **d. Elimination.**

28–31 The following is how Figure 14-2 should be labeled:

28 **b. Kidney;** 29 **c. Ureter;** 30 **a. Bladder;** 31 **d. Urethra.**

32 The answer is **a. Filtration.**

33 The answer is **b. Reabsorption** and **c. Secretion.**

34 The answer is **b. Reabsorption.**

35 The answer is **b. Reabsorption** and **c. Secretion.**

36 The answer is **b. Reabsorption** and **d. Excretion.**

Chapter 15

Fighting Enemies with the Immune System

You encounter bacteria and viruses all the time, some of which have the potential to make you very sick. Whether these potential pathogens cause you harm depends on a complicated give and take between their invasion tools and your defenses.

You emerge the winner from the vast majority of your microbial encounters because of the combination of your *innate immunity* (a built-in immune system that all humans have) and your *adaptive immunity* (the part of your immune system that develops as you encounter microbes). In this chapter, I introduce you to viral infections and present the structures and cells that keep you safe from microbes.

Microbial Encounters of the Best and Worst Kinds

Microbes are things like bacteria and viruses that are too small to see with the naked eye. They exist on every surface and in every environment on Earth. They're in the air, in the water, in the soil — even in your body. Most microbes can't hurt you, and many of them are beneficial to the environment or your body. But a few, called *pathogens,* grab all the headlines because they're the ones that cause diseases in humans.

✔ **Most microbes on Earth are beneficial to life and to humans.** Microbes in the environment are nature's recyclers. They break down the molecules in dead organisms and make them available again to living things. Humans harness the power of microbes for the production of fermented food like wine and cheese and for industrial applications like paper production. The bacteria that normally live in and on your body are your *normal microbiota,* and they protect your health by making your body less vulnerable to pathogens and by producing vitamins that aid with digestion and blood clotting.

✔ **A small percentage of microbes cause infectious diseases in humans.** *Infectious diseases* are simply those diseases caused by things that can spread, like bacteria and viruses. To cause disease, microbes must be able to enter and colonize your body, overcome your immune system's defenses, and cause damage to your body.

Viruses are an example of a microbe that can cause infectious disease. Think of viruses as the pirates of the microbial world — they hijack host cells and take them over to get what they need to reproduce.

Viruses aren't made of cells. They're just tiny particles of genetic information protected by a protein coat.

Viruses are much smaller than bacteria — so small you can't even see them with a light microscope. They don't have cell walls, ribosomes for protein synthesis, or the ability to transfer energy to ATP (see Chapter 3 for more on ribosomes and cell walls and Chapter 4 for the scoop on ATP and energy transfer). Because they have so little of their own cellular components (such as ribosomes), viruses can't reproduce unless they enter a host cell, which is why scientists call viruses *obligate intracellular parasites.* When viruses do find a host, they reprogram the cell with their own genetic information and convert the cell into a tiny factory that makes lots of copies of the virus.

Figure 15-1 shows an HIV virus attacking a human cell as an example of how viruses reproduce. Here are the steps the virus takes:

1. **Attachment: The virus attaches its proteins to a cell's *receptor*.**

   Think of this like inserting a key into a door. If your key doesn't fit, you can't get in. In #2 in Figure 15-1, you can see the HIV virus attaching to a protein called *CD4* that's found on the surface of certain human white blood cells.

2. **Penetration: The virus inserts its nucleic acid into the cell, taking over the cell.**

   In #3 in Figure 15-1, you can see the viral genetic material of HIV entering the human cell. The viral nucleic acid reprograms the cell, turning it into a viral production factory.

3. **Biosynthesis: Instead of doing its job for the body, the cell starts making viral nucleic acid and proteins.**

   The cell even uses its own molecules and energy reserves (ATP) to produce the viral parts. In #6 in Figure 15-1, you can see more viral genetic material being made by the cell.

   Some viruses, like HIV, can also insert a copy of their genetic material into the host cell chromosomes. Viruses that can do this become almost invisible to the immune system and are almost impossible to get rid of. And when the host cells reproduce, they make a copy of the viral genetic material, along with their own DNA, increasing the number of infected cells in the host.

4. **Maturation: The viral components pull themselves together to form mature viruses.**

   In #7 in Figure 15-1, you can see a viral particle in the process of maturing as it begins to exit the host cell.

5. **Release: The host cells release viral particles to go wreak havoc in other cells in the host's body.**

The number of viruses that go on attack at this point can range from ten to tens of thousands, depending on the type of virus. In #10 in Figure 15-1, you can see completed viral particles exiting the human cell.

1. **Free Virus.**

2. **Binding and Fusion:** Virus binds to CD4 and coreceptor on host cell and then fuses with the cell.

3. **Penetration:** The viral capsid enters the cell and releases its contents into the cytoplasm.

4. **Reverse Transcription:** The enzyme reverse transcriptase converts the single-stranded viral RNA molecules into double-stranded DNA.

5. **Recombination:** The enzyme integrase combines the viral DNA into the host cell DNA.

6. **Transcription:** Viral DNA is transcribed and translated to produce long chains of viral protein.

7. **Assembly:** Sets of viral proteins come together.

8. **Budding:** Release of immature virus occurs as viral proteins push out of the host cell, wrapping themselves in a new envelope. The viral enzyme protease begins cutting the viral proteins.

9. **Release:** Immature virus breaks free of the host cell.

10. **Maturation:** The viral enzyme protease finishes cutting the viral proteins, and the proteins combine to complete the formation of the virus.

Figure 15-1: How the HIV virus attacks cells.

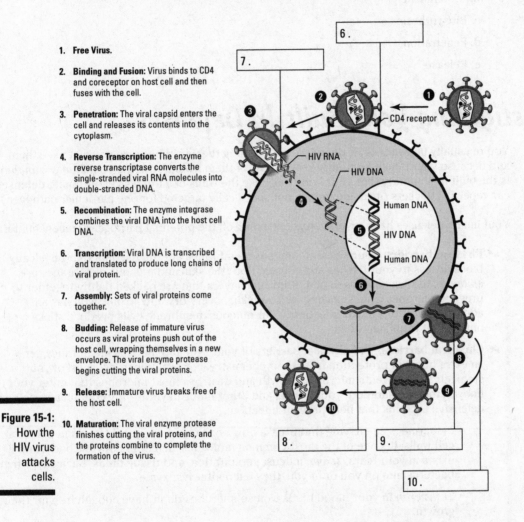

1.–5. Use the following terms to identify which type of pathogen best fits the characteristic in each question. Some questions may have more than one correct answer.

    a. Bacterial cell

    b. Virus

1. Has its own ribosomes.

2. Always requires a host cell in order to reproduce.

3. May insert its genetic material into the host chromosome.

4. Can cause infectious disease.

5. Could be killed by an antibiotic that targets cell walls.

6.–10. Use the following terms to label the basic steps in the viral life cycle shown in Figure 15-1.

   a. Maturation

   b. Attachment

   c. Biosynthesis

   d. Penetration

   e. Release

Investigating Your Built-In Defenses

You're usually unaware of all the microbes roaming the world because you can't see them and because your innate immunity keeps most of them from bothering you. *Innate immunity* is the built-in immunity that your body has. Like the walls of a fortress, your innate defenses can repel all attackers (meaning they're not specifically targeted for one particular pathogen).

Your innate defenses have several ways of fending off the potential pathogens you encounter:

✔ **Physical barriers:** Your skin and mucous membranes are the barriers that physically block access to your tissues and organs. Both the skin and mucous membranes are *epithelia,* tissues composed of multiple cell layers that are packed tightly together to prevent microbes from sneaking in. Your skin is very dry and strong, which is an additional barrier against infection. Your mucous membrane cells produce sticky mucus that traps microbes.

✔ **Chemical barriers:** The physical barriers of your skin and mucous membranes get an extra boost of protection from their chemistry. For example, the low pH of your stomach acid prevents microbial growth and destroys most microbes that enter your body in food. In addition, your mucus and other fluids in your body contain a variety of defensive proteins that help prevent infection:

   • *Lysozyme* is an enzyme that breaks down one of the chemicals found in bacterial cell walls. It's one of the most common molecular defenders in your body, and it exists in your tears, saliva, mucus, perspiration, and tissue fluids. Basically, when bacteria land on you or in you, they encounter lysozyme.

   • *Transferrin* in your blood binds iron so microbes don't have enough iron for their growth.

   • *Complement proteins* in your blood and tissue fluids bind to microbes and target them for destruction.

   • *Interferons* are released by cells infected with viruses. They travel to cells all around the infected cell and warn them about the virus. Cells that receive a warning from interferon produce proteins to help protect themselves against viral attack.

✔ **Dendritic cells:** These cells use special receptors, called *Toll-like receptors,* to recognize foreign molecules that make up microbial cells and alert your body. Dendritic cells activate your *adaptive immunity* (see the section "Fighting Back with Adaptive Immunity" later in this chapter) by breaking down bacteria and viruses and then presenting fragments of their molecules to other white blood cells, called *helper T cells.* Scientists call the molecules that can activate your immune system *antigens.*

✔ **Phagocytes:** Phagocytes like *macrophages* and *neutrophils* are white blood cells that seek and destroy microbes that have successfully entered your body. They actually wrap around invading microbes and eat them alive. Like dendritic cells, phagocytes activate helper T cells by showing them molecules from the destroyed microbes.

✔ **Inflammation:** When microbes do manage to invade, the microbes and your own damaged cells trigger a cascade of events that leads to *inflammation,* a local defensive response to cellular damage characterized by redness, pain, heat, and swelling. Inflammation fights infection by destroying microbes, confining the infection to one location, and repairing damaged tissue. Molecules such as histamine that are released during inflammation lead to vasodilation and increased blood vessel permeability.

- *Vasodilation* causes blood vessels to widen, allowing more blood to flow to the affected area to deliver clotting elements and white blood cells.

- *Increased blood vessel permeability* means the blood vessel walls loosen up, allowing white blood cells and materials to leave the blood and enter the tissues.

✔ **Filters:** The mucus in your nose and throat and the hairs in your nose act as filters that trap microbes and prevent them from getting deeper into your body. In the respiratory tract, a blanket of cilia called the *ciliary escalator* moves mucus upward in the throat to a location where you can cough it out, protecting your lower respiratory tract from infection (see Chapter 3 for more on cilia).

11.–19. For each of the following questions, name the component of your innate defenses that's the best match for the given description.

11. This cell engulfs and destroys pathogens.

12. This layer is one of your primary barriers to infection and protects you by trapping microbes in a sticky material.

13. This protein breaks down molecules found in bacterial cell walls.

14. This protein helps cells prepare themselves to fight off a viral attack.

15. This response helps limit the spread of infection in the body.

16. This protein binds iron in the body so that bacteria can't use it for their own growth.

17. The low pH of this fluid protects you from microbes trapped in your food.

18. This process widens blood vessels, allowing more blood to flow to a site of infection.

19. Having these in your nose helps prevent pathogens from getting into your respiratory system.

Fighting Back with Adaptive Immunity

Your body's innate defenses are incredible, and they prevent infection by most of the microbes that you encounter in your life. But every now and then, a microbe comes along that gets around your innate defenses and into your body. When your innate defenses are breached, it's time for the troops of your adaptive immunity to rally and fight back.

Your *adaptive immunity* gets its name because it adapts and changes as you go through life and are exposed to specific microbes that your innate defenses can't fight. If, for example, you're infected with *E. coli,* only those white blood cells that recognize particular molecules on *E. coli* are activated. If you face a different infection, say the bacteria *Staphylococcus aureus,* only the white blood cells that recognize specific molecules on *S. aureus* are activated. In other words, when your adaptive defenses come to your rescue, your body activates exactly the right team of white blood cells to fight each pathogen. That means your adaptive defenses learn to recognize specific pathogens after you encounter them.

One of the awesome features of your adaptive immunity is that it can remember a pathogen it has encountered before. This *immunologic memory* allows your immune system to respond much more effectively when you reencounter a particular pathogen.

Certain cells of your immune system, called *memory cells,* remain in a semiactivated state after your first encounter with a microbe. These memory cells and their descendants hang around for a long time after they're activated in the first battle. When the same pathogen shows up again, these cells multiply quickly and efficiently destroy the pathogen before you even realize it came back. Memory cells are therefore the reason why you can get some illnesses only once.

Several types of white blood cells work together to create your adaptive immunity:

- ✔ **Helper T cells:** Also called *CD4 cells,* these cells coordinate your entire adaptive immune response. Helper T cells receive signals from the white blood cells of your innate defenses, such as dendritic cells and phagocytes, and relay those signals to the fighters of your adaptive defenses: the B cells and cytotoxic T cells.

- ✔ **B cells:** These cells are activated when they detect a foreign pathogen with their B cell receptors or when they receive signals from helper T cells. They're activated to form two types of cells: plasma cells and memory cells.

Plasma cells produce *antibodies,* defensive proteins that bind specifically to antigens. Your immune system releases the antibodies that plasma cells produce into the blood, where they can circulate around the body. Anything in the body that's tagged with antibodies — such as invading pathogens — is marked for destruction by the immune system.

- ✔ **Cytotoxic T cells:** Also called *CD8 cells* or *cytotoxic T lymphocytes (CTLs),* these cells come into play if microbes try to hide inside your cells so that the antibodies can't find them. Cytotoxic T cells can detect foreign antigens on the surface of an infected host cell. When these cells discover an infected cell, they send signals that tell the infected cell to commit suicide — a necessary sacrifice in order to destroy the hidden microbes.

Of all these types of white blood cells, your helper T cells are probably the most important. *Antigen-presenting cells* like dendritic cells and macrophages from your innate immunity activate helper T cells by showing them bits of molecules from pathogens. After they're activated, your helper T cells multiply and release communicating molecules called *cytokines* that stimulate both cytotoxic T cells and B cells. Thus, without the action of helper T cells, your entire immune system would fail.

The HIV virus infects helper T cells, slowly reducing their numbers until a person who's infected with the virus doesn't have enough helper T cells to activate his adaptive immunity. At this point, the person develops *acquired immunodeficiency syndrome,* better known as AIDS. After a person has AIDS, he's very susceptible to infection and certain cancers, which ultimately cause the person's death.

The activation of helper T cells and the other cells that make up your immune system involves several steps:

1. **Antigen-presenting cells attach pieces of the foreign antigen to proteins that they display on their surface.**

2. **Antigen-presenting cells also produce molecules like cytokines, signaling that they've detected a foreign antigen.**

3. **Helper T cells bind to the displayed antigen using a receptor called a *T cell receptor.***

4. **After helper T cells recognize antigen and receive the signals from antigen-presenting cells, they activate; activated helper T cells multiply and then activate cytotoxic T cells and B cells.**

20.–30. For each of the following questions, name the component of your adaptive defenses that's the best match for the given description.

20. These cells kill cells infected with viruses.

21. These proteins stick to foreign molecules, marking them for destruction by the immune system.

22. These cells produce signals that activate cytotoxic T cells and B cells.

23. These cells produce antibodies.

24. These cells have a protein on their surface called CD8.

25. These cells are the host cell for the HIV virus.

26. Cells use these molecules to communicate with one another.

27. These cells can become plasma cells and memory cells.

28. These cells show antigens to helper T cells.

29. These cells live a long time and help you respond quickly to pathogens when you encounter them for a second time.

30. These molecules enter the body as part of pathogens, triggering your adaptive immune response.

31.–38. Use the terms that follow to label the cells and steps that occur during activation of your immune system in Figure 15-2.

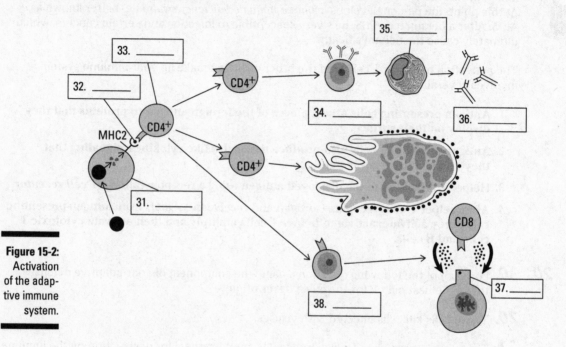

Figure 15-2:
Activation of the adaptive immune system.

a. B cell

b. Cytotoxic T cell

c. Antigen

d. Antibody

e. Antigen-presenting cell

f. Helper T cell

g. Infected cell

h. Plasma cell

Answers to Questions on the Immune System

The following are answers to the practice questions presented in this chapter.

1 The answer is **a. Bacterial cell.**

2 The answer is **b. Virus.**

3 The answer is **b. Virus.**

4 The answers are **a. Bacterial cell** and **b. Virus.**

5 The answer is **a. Bacterial cell.**

Viruses aren't cells, so they don't have cellular structures like cell walls. Because antibiotics target cellular structures, they don't work on viruses.

6 The answer is **b. Attachment.**

7 The answer is **d. Penetration.**

8 The answer is **c. Biosynthesis.**

9 The answer is **a. Maturation.**

10 The answer is **e. Release.**

11 The answer is **phagocyte** (or neutrophil or macrophage).

12 The answer is **mucous membrane.**

13 The answer is **lysozyme.**

14 The answer is **interferon.**

15 The answer is **inflammation.**

16 The answer is **transferrin.**

17 The answer is **stomach acid.**

18 The answer is **vasodilation.**

19 The answer is **hairs** (or mucus).

20 The answer is **cytotoxic T cells.**

21 The answer is **antibodies.**

22 The answer is **helper T cells.**

23 The answer is **plasma cells.**

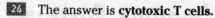

24 The answer is **cytotoxic T cells.**

25 The answer is **helper T cells.**

26 The answer is **cytokines.**

27 The answer is **B cells.**

28 The answer is **antigen-presenting cells** (or dendritic cells or macrophages).

29 The answer is **memory cells.**

30 The answer is **antigens.**

31 – 38 The following is how Figure 15-2 should be labeled:

31 **e. Antigen-presenting cell;** 32 **c. Antigen;** 33 **f. Helper T cell;** 34 **a. B cell;** 35 **h. Plasma cell;** 36 **d. Antibody;** 37 **g. Infected cell;** 38 **b. Cytotoxic T cell.**

Chapter 16

Sending Messages with the Nervous and Endocrine Systems

In This Chapter

▶ Looking at the parts and functions of the nervous system

▶ Differentiating among different kinds of neurons

▶ Following the path of a nerve impulse

▶ Homing in on hormones

With all the metabolic processes and reactions going on in living things, organisms need to be able to exert some control in order to avoid chaos. Enter the nervous and endocrine systems. These systems regulate your body's responses to what it encounters to help you maintain *homeostasis,* the internal balance of the body. In this chapter, I present the basic structure and function of both systems and explore how the human body responds to their signals.

Mapping out Nervous Systems

Animals are the only living things on Earth with complex nervous systems that first receive and interpret sensory signals from the environment and then send out messages to direct the animal's response. The complexity of an animal's nervous system depends on its lifestyle and body plan.

✔ Animals whose bodies don't have a defined head or tail have *nerve nets,* which are weblike arrangements of nerve cells that extend throughout the body.

✔ Animals with a defined head possess a two-part nervous system:

- The *central nervous system* (CNS) consists of the animal's brain and central neurons. It's housed in the head and may continue along the back.

- The *peripheral nervous system* (PNS) consists of all the nerves that travel from the CNS to the rest of the animal's body.

In all animals with a backbone, including you, the CNS consists of a brain and a spinal cord. The *brain* contains centers that process information from the sense organs, centers that control emotions and intelligence, and centers that regulate homeostasis. The *spinal cord* controls the flow of information to and from the brain.

Both the brain and the spinal cord are highly protected. First of all, they sit within a liquid called *cerebrospinal fluid* that guards the CNS against shocks caused by movement, helps supply nutrients, and helps remove wastes. The blood-brain barrier, which is created by the capillaries surrounding the brain, provides another layer of protection because the capillaries are highly selective about what they allow to enter the brain or cerebrospinal fluid. A final layer of protection is the *meninges,* two layers of connective tissue that surround the brain and spinal cord.

From the CNS, the nervous system branches off into the PNS, which is divided into two systems:

- **Somatic nervous system:** This part of the PNS carries signals to and from the skeletal muscles. It controls many of an animal's voluntary responses to signals in its environment.

- **Autonomic nervous system:** This part of the PNS controls the mostly involuntary internal processes in the body, such as heartbeat and digestion. It has two divisions that work opposite each other to maintain homeostasis:

  - The *sympathetic nervous system* automatically stimulates the body when action is required. This is the part of the nervous system responsible for the *fight-or-flight response,* which stimulates a surge of adrenaline to give the body quick energy so it can escape danger. The sympathetic nervous system also quickens the heart rate to move blood through the blood vessels faster and releases sugar from the liver's glycogen stores into the blood so fuel is readily available to the cells.

  - The *parasympathetic nervous system* stimulates more routine functions, such as the secretion of digestive enzymes or saliva. In contrast to the sympathetic nervous system, the parasympathetic nervous system slows down the heart rate after the fight-or-flight response is no longer needed.

1. Use two different colored pencils or highlighters to mark Figure 16-1. With one color, highlight the CNS. With the other color, highlight the PNS.

2.–4. Use the following terms to identify the type of nervous system that would be involved in each example.

    a. Somatic nervous system

    b. Sympathetic nervous system

    c. Parasympathetic nervous system

2. You're about to give a presentation in front of a class. Your heart is pounding and you feel a little lightheaded.

3. You reach into a refrigerator to get yourself a drink.

4. The heart rate of a black bear slows as it settles into dormancy.

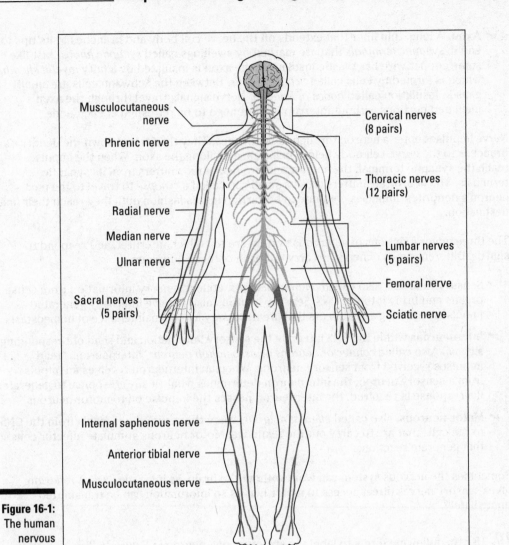

Musculocutaneous nerve

Phrenic nerve

Radial nerve

Median nerve

Ulnar nerve

Sacral nerves
(5 pairs)

Cervical nerves
(8 pairs)

Thoracic nerves
(12 pairs)

Lumbar nerves
(5 pairs)

Femoral nerve

Sciatic nerve

Internal saphenous nerve

Anterior tibial nerve

Musculocutaneous nerve

Figure 16-1:
The human
nervous
system.

From LifeART®, Super Anatomy 1, © 2002, Lippincott Williams & Wilkins

Getting on Your Nerves

Neurons are cells that form the core of nervous systems because they have the ability to receive and transmit signals. Neurons have a unique elongated shape and consist of three main parts:

- ✔ **Nerve cell body:** The rounded part of the neuron. It contains typical eukaryotic cell components like the nucleus, organelles, and the endomembrane system (see Chapter 3 for more on cells).

- ✔ **Dendrites:** Tiny projections that branch off the nerve cell body at the neuron's receiving end. The dendrites act like tiny antennae that pick up signals from other cells.

✔ **Axon:** A long, thin fiber that extends off the nerve cell body and branches at its tips to end in *synaptic terminals* that are marked by swellings called *synaptic knobs*. Just like some copper wire has plastic insulation, the axon is insulated by a fatty *myelin sheath*, which is formed by cells called *Schwann cells*. Between the Schwann cells are small gaps in insulation called *nodes of Ranvier*. Nerve signals travel through the axon's insulated portions without interruption but need to be refreshed at each node.

Nerve impulses enter a neuron through the dendrites. They then travel down the dendrite's branches to the nerve cell body before being carried along the axon. When the impulses reach the synaptic terminal, the neuron releases neurotransmitters from its synaptic terminals. The neurotransmitters cross a small gap called a *synapse* to travel to the next neuron's dendrites. Impulses continue to be carried in this fashion until they reach their final destination.

The three major functions of a nervous system are to collect, interpret, and respond to signals. Different types of neurons carry out each of these functions.

✔ **Sensory neurons,** also called *afferent neurons,* collect sensory information from sense organs and bring it to the CNS. Sensory neurons also receive internally generated impulses regarding adjustments that are necessary for the maintenance of homeostasis.

✔ **Interneurons** within the CNS integrate the sensory information and send out responding signals. Also called *connector neurons* or *association neurons,* interneurons "read" impulses received from sensory neurons. When an interneuron receives an impulse from a sensory neuron, the interneuron determines what (if any) response to generate. If a response is required, the interneuron passes the impulse on to motor neurons.

✔ **Motor neurons,** also called *efferent neurons,* carry the responding signals from the CNS to the cells that are to carry out the response. Motor neurons stimulate effector cells that generate reactions.

Sometimes the nervous system can work without the brain, as in a reflex arc. A *reflex arc* gives sensory nerves direct access to motor nerves so information can be transmitted immediately.

5.–10. Use the following terms to label the structure of a neuron in Figure 16-2.

    a. Node of Ranvier

    b. Schwann cell

    c. Dendrite

    d. Cell body

    e. Axon

    f. Synaptic terminal

11. Use four different colored pencils or highlighters to highlight Figure 16-2. Color the portion of the neuron that receives the signal one color. Color the cell body another color. Color the portion of the neuron that transmits the signal a third color. Finally, color the insulation a fourth color.

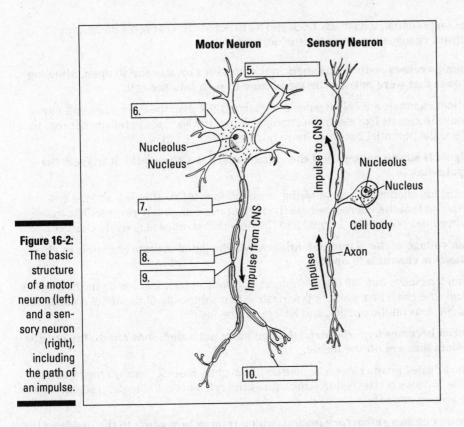

Figure 16-2:
The basic
structure
of a motor
neuron (left)
and a sen-
sory neuron
(right),
including
the path of
an impulse.

Getting in on the Action Potential

When a neuron is inactive, just waiting for a nerve impulse to come along, the neuron is *polarized* — that is, the cytoplasm inside the cell has a negative electrical charge, and the fluid outside the cell has a positive charge. This separation of charge sets up conditions for the neuron to respond, just like a separation of charge in a battery sets up conditions that allow a battery to provide electricity.

The electrical difference across the membrane of the neuron is called its *resting potential*.

The resting potential is created by a transport protein called the *sodium-potassium pump*. This protein moves large numbers of sodium ions (Na^+) outside the cell, creating the positive charge. At the same time, the protein moves some potassium (K^+) ions into the cell's cytoplasm. Because the number of Na^+ ions moved outside the cell is greater than the number of K^+ ions moved inside, the cell is more positive outside than inside.

When a stimulus reaches a resting neuron, the neuron transmits the signal as an impulse called an *action potential*.

During an action potential, ions cross back and forth across the neuron's membrane, causing electrical changes that transmit the nerve impulse:

1. **The stimulus causes sodium channels in the neuron's membrane to open, allowing the Na⁺ ions that were outside the membrane to rush into the cell.**

   The sodium channels are called *gated ion channels* because they can open and close in response to signals like electrical changes. When the Na⁺ ions enter the neuron, the cell's electrical potential becomes more positive.

2. **If the signal is strong enough and the voltage reaches a *threshold,* it triggers the action potential.**

   More gated ion channels open, allowing more Na⁺ ions inside the cell, and the cell *depolarizes* so that the charges across the membrane completely reverse: The inside of the cell becomes positively charged and the outside becomes negatively charged.

3. **The peak voltage of the action potential causes the gated sodium channels to close and potassium channels to open.**

   Potassium ions move outside the membrane, and sodium ions stay inside the membrane, *repolarizing* the cell. The result is a polarization that's opposite of the initial polarization that had Na⁺ ions on the outside and K⁺ ions on the inside.

4. **The neuron becomes *hyperpolarized* when more potassium ions are on the outside than sodium ions are on the inside.**

   When the K⁺ gates finally close, the neuron has slightly more K⁺ ions on the outside than it has Na⁺ ions on the inside. This causes the cell's potential to drop slightly lower than the resting potential.

5. **The neuron enters a *refractory period,* which returns potassium to the inside of the cell and sodium to the outside of the cell.**

   The sodium-potassium pump goes back to work, moving Na⁺ ions to the outside of the cell and K⁺ ions to the inside, returning the neuron to its normal polarized state.

12.–17. Use the terms that follow to label the action potential in Figure 16-3.

    a. Threshold

    b. Resting potential

    c. Depolarization

    d. Repolarization

    e. Hyperpolarization

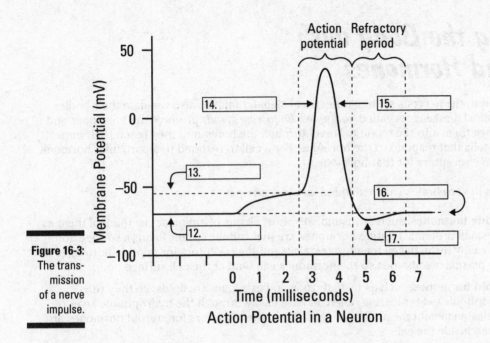

Figure 16-3:
The trans-
mission
of a nerve
impulse.

18.–21. Use the following terms to fill in the blanks in each statement.

    a. Inside

    b. Outside

    c. Positively charged

    d. Negatively charged

18. The sodium-potassium pump moves sodium to the _____ of the cell.

19. The sodium-potassium pump moves potassium to the _____ of the cell.

20. During a resting potential, the cell's cytoplasm is _____ relative to the out-side of the cell.

21. At the peak of the action potential, the cell's cytoplasm is _____ relative to the outside of the cell.

Regulating the Body with Glands and Hormones

In addition to the nervous system's electrical signals, animals also regulate their bodies with chemical messengers called *hormones*. *Endocrine glands* produce the hormones and then release them into the blood to travel through the body until they reach their *target cells,* the cells that respond to the hormone. For a cell to respond to a particular hormone, it must have receptors for that hormone.

Hormones in vertebrates can be divided into two groups:

- ✔ **Peptide hormones,** such as insulin, are short chains of amino acids; think of them as very small proteins. Peptide hormones are hydrophilic (water-loving), so they don't pass easily through cell membranes. Cells put the receptors for peptide hormones in their plasma membranes so the hormones can bind at the cell surface.

- ✔ **Steroid hormones,** such as testosterone and estrogen, are lipids, so they're hydrophobic (water-fearing) and can pass easily through the hydrophobic layer of the plasma membrane and enter cells. Thus, the receptors for steroid hormones are located inside the cell.

When a signal like a hormone reaches a cell and is relayed to molecules within the cell, scientists call it *signal transduction.*

Hormones trigger cellular responses in three basic steps:

1. **Reception: The hormone binds to its receptor.**

    Hydrophilic hormones typically bind to receptors on the cell surface, whereas hydrophobic hormones can pass through the plasma membrane and bind to a receptor inside the cell.

2. **Signal transduction: When the hormone binds to its receptor, it causes a change in the receptor that is passed to molecules inside the cell.**

    After the receptor changes, it causes changes in another molecule, which causes changes in another molecule, and so on. Scientists call the molecules in this signaling chain *relay molecules* or *second messengers.*

3. **Cellular response: A final member of the relay molecules causes a change in the cell's behavior.**

    The relay molecule may cause the cell to access a gene to build a new protein or to stop making a protein. Or the relay molecule may interact with an enzyme to increase or stop its activity.

The *endocrine system* is the system that handles hormone production and secretion within an organism. It keeps a check on cellular processes and the bloodstream's components and can make adjustments as necessary.

22.–27. Use the following terms to label the events of transduction of a hormone signal shown in Figure 16-4.

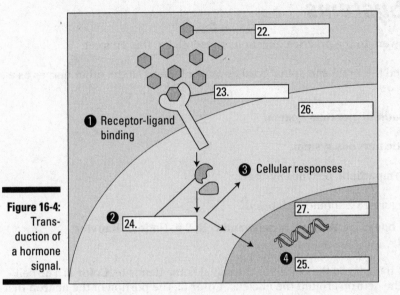

Figure 16-4:
Trans-
duction of
a hormone
signal.

a. Hormone (primary messenger)

b. Relay proteins (secondary messengers)

c. DNA

d. Cytoplasm

e. Nucleus

f. Receptor

28. Use different colored pencils or highlighters to mark the sections of Figure 16-4 that show the major events that occur during the transduction of a hormone signal:

a. Reception of hormone signal

b. Signal transduction

c. Response by changes in cellular activity

Answers to Questions on the Nervous and Endocrine Systems

The following are answers to the practice questions presented in this chapter.

1 You should have colored the brain and spinal cord as one color and all the other nerves as a different color.

2 The answer is **b. Sympathetic nervous system.**

3 The answer is **a. Somatic nervous system.**

4 The answer is **c. Parasympathetic nervous system.**

5–10 The following is how Figure 16-2 should be labeled:

5 c. Dendrite; 6 d. Cell body; 7 e. Axon; 8 b. Schwann cell; 9 a. Node of Ranvier; 10 f. Synaptic terminal

11 **Color 1:** The portion of the neuron that receives the signal is the dendrite. **Color 2:** The cell body is the portion of the neuron around the nucleus. **Color 3:** The portion of the neuron that transmits the signal is the axon. **Color 4:** The Schwann cells provide the insulation (myelin sheath).

12–17 The following is how Figure 16-3 should be labeled:

12 b. Resting potential; 13 a. Threshold; 14 c. Depolarization; 15 d. Repolarization; 16 b. Resting potential; 17 e. Hyperpolarization

18 The answer is **b. Outside.**

19 The answer is **a. Inside.**

20 The answer is **d. Negatively charged.**

21 The answer is **c. Positively charged.**

22–27 The following is how Figure 16-4 should be labeled:

22 a. Hormone (primary messenger); 23 f. Receptor; 24 b. Relay proteins (secondary messengers); 25 c. DNA; 26 d. Cytoplasm; 27 e. Nucleus

28 **a. Reception of hormone signal** should highlight the hormone and the receptor. **b. Signal transduction** should highlight everything from the inside edge of the receptor through the relay proteins up to the DNA. **c. Response by changes in cellular activity** should highlight the arrow that moves off the side of the relay proteins.

Chapter 17

Making Babies with the Reproductive System

*I*n some ways, human reproduction is simple — sperm meets egg, and a baby is made. But a lot has to happen to get to that big event. For one thing, meiosis (see Chapter 5) occurs in the gonads of both parents to produce gametes in a process called *gametogenesis*. Those gametes must make their way through the reproductive systems of both parents to be in just the right spot when they find each other. And when the gametes do join together, the mother's body must be ready to support and nourish a developing embryo. In this chapter, I introduce you to the basic structure and function of the human reproductive system and then show you the very earliest events that occur in every human.

Identifying the Parts of the Male Reproductive System

Before you dive into the details of human sexual reproduction, knowing a little bit about the organ systems involved helps. Figure 17-1 illustrates the male reproductive system:

✔ The egg-shaped male gonads, called *testes,* rest in a sac called the *scrotum.* Sperm don't develop normally at the human body's core temperature, so keeping them cooler outside the body allows normal development.

✔ The staff-like *penis* consists of erectile tissue (shaded in Figure 17-1) that surrounds a narrow tube called the *urethra* (part of the excretory system; see Chapter 14 for details). During sexual intercourse, the erectile tissue fills with blood, allowing the penis to stiffen so that it can enter the female's vagina. The head of the penis, called the *glans,* contains many nerve endings and is very sensitive to stimulation. At birth, a flap of skin called the *prepuce* (or *foreskin*) covers the glans. In some cultures, the foreskin is surgically removed by *circumcision.*

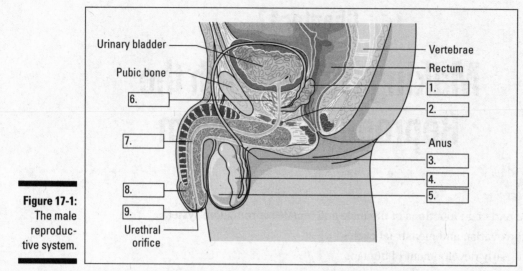

Figure 17-1:
The male
reproduc-
tive system.

From LifeART®, Super Anatomy 1, © 2002, Lippincott Williams & Wilkins

Sperm production, called *spermatogenesis,* begins in each testis. The sperm mature as they follow this path out of the male reproductive system:

1. **Sperm move into the *epididymus,* a coiled tube that rests on the testis.**

   The epididymus stores the sperm as they continue to develop.

2. **During *ejaculation,* when sperm-containing fluid leaves the penis, muscular contractions push sperm from the epididymus into a tube called the *vas deferens.***

3. **The sperm travel through the vas deferens up into the abdomen and around the bladder until they reach a gland called the *seminal vesicle.***

   During ejaculation, the seminal vesicle secretes a thick solution that provides the sperm with fructose to use as a source of energy as they swim through the female reproductive system.

4. **The vas deferens and seminal vesicle join to form a short duct called the *ejaculatory duct.***

   The ejaculatory duct passes through the *prostate gland,* which secretes a thin fluid that provides nutrients to the sperm.

5. **The ejaculatory ducts from both sides of the male reproductive system join and empty into the *urethra,* which carries the sperm out of the body during sexual intercourse.**

   Mucus from the *bulbourethral gland,* which joins the urethra just after the ejaculatory duct, also empties into the urethra. *Semen,* the fluid released during ejaculation, consists of sperm plus the fluids from the three glands in the male reproductive system.

When the pleasurable feelings of *orgasm* occur in a male, a sphincter muscle closes off the bladder to prevent urine from entering the urethra. Shutting out urine allows the urethra to be used solely for ejaculation at that time.

1.–9. Use the terms that follow to label the structures of the male reproductive system in Figure 17-1.

   a. Testes

   b. Vas deferens

   c. Prepuce (foreskin)

   d. Seminal vesicle

   e. Epididymis

   f. Urethra

   g. Prostate gland

   h. Bulbourethral gland

   i. Glans penis

10. Put the terms that follow in order to show the path of sperm from the testes out of the male reproductive system.

   a. Ejaculatory duct

   b. Testes

   c. Vas deferens

   d. Urethra

   e. Epididymis

Identifying the Parts of the Female Reproductive System

The sperm is only half the equation; a sperm must fertilize an egg to start the reproductive process. The female reproductive system produces eggs and supports the developing fetus. Figure 17-2 illustrates the structures of the female reproductive system.

✔ The female gonads are oval, lumpy-looking structures called *ovaries*. Ovaries produce eggs in a process called *oogenesis*. The lumps are the *follicles*, layers of cells that surround and protect the developing egg cells. In addition to producing egg cells, the ovaries produce the hormone *estrogen*.

✔ During *ovulation*, an egg cell is pushed out of the ovary into a *fallopian tube*, which carries the egg to the uterus.

✔ The *uterus* is a muscular structure that houses and protects a developing fetus. Its lining, the *endometrium*, contains lots of blood vessels that bring nutrients and oxygen to the fetus.

✔ The uterus narrows toward the outside of the body, forming a dome of tissue called the *cervix*. The cervix forms at the junction between the uterus and the muscular canal called the *vagina*. The vagina opens to the outside of the body, just behind the opening of the urethra (part of the excretory system; see Chapter 14 for details).

✔ Just in front of the urethra is a small bud of tissue called the *clitoris*. The clitoris consists of internal tissue called the *glans* that's covered with a small hood of tissue called the *prepuce*. Just like the glans of the penis, the glans clitoris is packed with nerve endings and is very sensitive to touch stimulation. It also contains erectile tissue and swells with blood during sexual arousal.

✔ Two sets of tissue fold over and protect the female reproductive parts. A thin pair of tissue folds, called the *labia minora,* forms a border around the vaginal opening. The *labia majora,* a thicker, fatty pair of folds, forms a protective covering over the labia minora and vaginal opening.

In females, the height of sexual stimulation also causes intense muscular contractions and a pleasurable feeling of release called *orgasm*. The fluid released inside the vagina helps create a watery environment that the sperm can swim in. The uterus's muscular contractions slightly open the cervix, which allows sperm to get inside the uterus and also helps pull sperm upward toward the fallopian tubes.

11.–18. Use the terms that follow to label the parts of the female reproductive system in Figure 17-2.

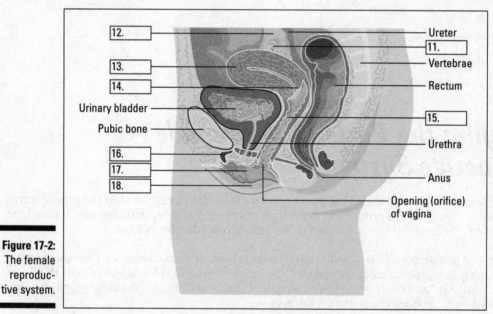

Figure 17-2: The female reproductive system.

From LifeART®, Super Anatomy 1, © 2002, Lippincott Williams & Wilkins

a. Vagina

b. Clitoris

c. Uterus

d. Ovary

e. Fallopian tube

f. Labia majora

g. Labia minora

h. Cervix

Following the Female Ovarian and Menstrual Cycles

Oogenesis (the egg-producing process) begins very early in the life of a human female, when she's still a developing fetus! In fact, a human female is born with all the eggs she'll ever have. However, the eggs aren't quite finished, because cell division pauses early in the first half of meiosis (see Chapter 5 for more on meiosis). These eggs lie dormant from birth until puberty, at which time the monthly cycle of a woman's hormones restarts the development of eggs — usually one per month.

Thus, two reproductive cycles in the female regulate human sexual reproduction (see Figure 17-3):

✔ **The monthly *ovarian cycle* regulates development of the egg in the ovary.** The ovarian cycle includes the development of the follicle, the secretion of hormones by the follicle, ovulation, and the formation of the corpus luteum. It occurs in the ovary, takes about 28 days, and is controlled by the hormones GnRH, FSH, LH, and estrogen.

✔ **The monthly *menstrual cycle* refers to the periodic series of changes that prepare the female body for the implantation of an embryo.** The menstrual cycle includes the thickening of the endometrium to prepare for possible implantation of an embryo and the shedding of the endometrium if no embryo is implanted. It occurs in the uterus, takes about 28 days, and is controlled by the levels of the hormones *progesterone* and *estrogen*.

Ultimately, the brain controls the ovarian cycle by releasing hormones from two glands at its center. The *hypothalamus* monitors the levels of estrogen and progesterone in the blood. When the levels decline, the hypothalamus secretes a hormone called *gonadotropin-releasing hormone* (GnRH), which goes straight to the *pituitary gland* and stimulates part of the pituitary to secrete *follicle-stimulating hormone* (FSH) and *leuteinizing hormone* (LH).

The hormones from the brain control the levels of estrogen and progesterone released by the female reproductive system, leading to the events of the ovarian cycle:

1. **A follicle starts to grow and begins producing the hormone *estrogen.***

2. **About midway through the ovarian cycle, estrogen reaches a critical level, causing the hypothalamus to release more GnRH.**

   The GnRH triggers the pituitary gland to release a burst of LH and FSH.

3. **The spike in LH triggers the completion of meiosis so that the egg finishes its development.**

   LH stimulates enzymes that break open the follicle, causing *ovulation,* the release of the egg from the follicle.

4. **LH also triggers the remaining follicle cells to develop into a mass of cells called the *corpus luteum,* which secretes estrogen and progesterone for the rest of the ovarian cycle (about two more weeks).**

5. **Estrogen and progesterone prepare the body for a possible pregnancy by spurring the tissues lining the uterus to develop thicker blood vessels.**

   These hormones send negative feedback to the hypothalamus and pituitary, reducing the production of FSH and LH and thus preventing the development of more follicles.

6. The corpus luteum breaks down after about two weeks, and if implantation of a fertilized egg doesn't occur, the levels of estrogen and progesterone fall, and the entire ovarian cycle begins again.

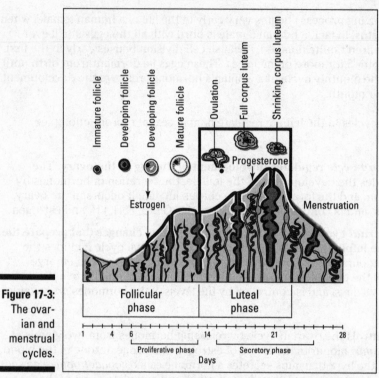

Figure 17-3:
The ovarian and menstrual cycles.

From LifeART®, Super Anatomy 1, © 2002, Lippincott Williams & Wilkins

The progesterone and estrogen from the ovarian cycle also control the menstrual cycle:

- ✔ If implantation doesn't occur and the levels of estrogen and progesterone fall, the endometrium sloughs off, and menstrual bleeding begins.

- ✔ As the levels of estrogen begin to rise, the endometrium thickens. This thickening continues as progesterone levels increase.

The first day of menstrual bleeding marks the first day of the menstrual cycle.

If a fertilized egg does implant in the uterus's lining, the developing embryo releases a hormone called *human chorionic gonadotropin* (hCG). The presence of hCG maintains the corpus luteum, thus keeping up production of the estrogen and progesterone necessary to maintain the endometrium. After the *placenta* (a blood-filled, nutrient-rich, temporary organ) has formed, the embryo gets its nutrients and blood supply through the umbilical cord connecting the embryo to the placenta, which is connected to the mother's blood supply. Therefore, the embryo's production of hCG declines after the placenta is up and running.

19.–24. Use the terms that follow to identify the role of each hormone. Some questions may have more than one correct answer.

    a. FSH

    b. LH

    c. hCG

    d. GnRH

    e. Estrogen

    f. Progesterone

19. Falling levels of this trigger the start of menstruation.

20. It stimulates the pituitary to make FSH and LH.

21. It triggers growth of the follicle.

22. It causes rupture of the follicle and ovulation.

23. Rising levels of this stimulate development of the endometrium.

24. It prevents breakdown of the corpus luteum.

Fertilization through Birth: Developing New Humans

Fertilization, the joining of sperm and egg, typically occurs in the fallopian tube, after sperm have made the long swim up through the vagina, past the cervix, and through the uterus. Fertilization brings together the chromosomes from each parent, creating the first cell, or *zygote,* of the new human.

Because a human egg lives no longer than 24 hours after ovulation and human sperm live no longer than 72 hours, intercourse that occurs in the three-day period prior to ovulation or within the day after ovulation is the only chance of fertilization during a given month.

The zygote divides by mitosis, beginning production of all the cells necessary for the human body. The development of a new organism occurs through the production and specialization of new cells. The development from zygote to newborn occurs in several stages.

Embryonic development occurs from the zygote through the eighth week of pregnancy. The zygote begins traveling down the fallopian tube, heading for the uterus so it can implant in the uterine lining. As it travels, the zygote undergoes *cleavage,* a rapid series of mitotic divisions that result in a multicellular embryo.

✔ After cell division produces a solid ball of 16 cells, the developing human is called a *morula.*

✔ The morula fills with liquid, forming a hollow ball of cells called a *blastula.*

✔ In humans, a group of cells inside the blastula becomes specialized to form the *embryo,* and the blastula becomes a *blastocyst.* Different layers of cells become specialized within the blastocyst, taking the first step toward forming specialized tissues. (You can recognize the blastocyst by the flattened cells along its edge, called the *trophoblast.*)

Conception occurs when the blastocyst successfully implants itself in the uterine wall.

Don't make the mistake of thinking that conception is the same as fertilization. The egg can be fertilized, but a woman isn't pregnant until the blastocyst is rooted in her uterine wall, where it can develop further.

After implanting itself, the developing mass of cells moves inward, forming a ball of cells called a *gastrula* that has three layers of cells; this process is referred to as *gastrulation.* Each layer of cells in the gastrula eventually becomes a different type of tissue:

✔ The outer layer is the *ectoderm,* which develops into the skin and nervous systems.

✔ The middle layer is the *mesoderm,* which develops into the muscular, skeletal, and circulatory systems.

✔ The innermost layer is the *endoderm,* which gives rise to the linings of the digestive and respiratory tracts, as well as organs such as the liver and pancreas.

After gastrulation, the specialized cells of the ectoderm, mesoderm, and endoderm begin migrating toward other cells with the same specialty. This cellular migration is referred to as *morphogenesis* because it gives the embryo a shape. As cells differentiate, organ systems begin to form. At the end of the embryonic period, a human embryo is about an inch and a half long and has recognizable human body structures.

While the embryo is developing, structures form outside the embryo that support and nourish the embryo and fetus. Two specialized membranes develop outside the embryo. One, called the *chorion,* combines with tissues created by the mother to become the *placenta.* The other, called the *amnion,* surrounds the amniotic cavity, which fills with *amniotic fluid* that protects the developing embryo and fetus.

A tubular structure called the *allantois* forms off of the center cavity of the blastocyst. In humans, the allantois eventually becomes the *body stalk* and then the *umbilical cord,* which connects the fetus to the placenta.

After the eighth week of pregnancy, the developing human is a *fetus.* Fetuses are completely differentiated, meaning their cells have migrated and formed organ systems. All fetuses do in the uterus is continue to grow and develop features such as hair and nails. As the fetus gets stronger, longer, and heavier, it looks more and more like a newborn baby.

During the last weeks of pregnancy, the hormone estrogen reaches high levels in the mother's blood. Estrogen triggers the formation of receptors for another hormone, called *oxytocin,* on the uterine wall. Labor is triggered when the fetus starts producing oxytocin, which causes the mother's pituitary to produce even more oxytocin. The uterus's receptors receive the oxytocin signals, causing uterine muscles to start labor *contractions.* Oxytocin also stimulates the placenta to produce *prostaglandins,* which increase uterine contractions.

Labor occurs in three stages:

1. *Dilation* **refers to the beginning of labor until the cervix has opened (dilated) to a diameter of 10 centimeters.**

   Dilation is the longest period of labor, typically lasting from 6 to 12 hours but sometimes quite a bit longer.

2. *Expulsion* **is the delivery of the infant.**

   The mother experiences strong uterine contractions and an urge to push. Expulsion usually lasts for 20 minutes to one hour.

3. **Delivery of the placenta usually occurs within 15 minutes of expulsion.**

When the fetus is finally born, the infant is called a *neonate,* meaning newborn. A life begins, and development continues.

25.–31. Use the terms that follow to identify the correct stage of a developing human.

   a. Blastula

   b. Fetus

   c. Morula

   d. Embryo

   e. Zygote

   f. Blastocyst

   g. Neonate

25. A fluid-filled ball of cells.

26. A multicellular human at four weeks.

27. A single cell formed from the fusion of sperm and egg.

28. A newborn baby.

29. A solid ball of 16 cells.

30. A developing human at five months.

31. A ball of cells containing a trophoblast and a developing embryo.

32.–35. Use the terms that follow to correctly identify the major events that occur during an embryo's development.

    a. Differentiation of organ systems

    b. Morphogenesis

    c. Cleavage

    d. Gastrulation

32. A ball of cells develops three specialized layers of cells.

33. Specialized cells of the same type migrate toward one another.

34. A zygote undergoes a rapid series of cell divisions.

35. Organs begin to form.

Answers to Questions on the Human Reproductive System

The following are answers to the practice questions presented in this chapter.

1–**9** The following is how Figure 17-1 should be labeled:

1 d. Seminal vesicle; 2 h. Bulbourethral gland; 3 b. Vas deferens; 4 e. Epididymis; 5 a. Testes; 6 g. Prostate gland; 7 f. Urethra; 8 i. Glans penis; 9 c. Prepuce (foreskin).

10 The answer is **b. Testes** → **e. Epididymis** → **c. Vas deferens** → **a. Ejaculatory duct** → **d. Urethra.**

11–**18** The following is how Figure 17-2 should be labeled:

11 d. Ovary; 12 e. Fallopian tube; 13 c. Uterus; 14 h. Cervix; 15 a. Vagina; 16 b. Clitoris; 17 g. Labia minora; 18 f. Labia majora.

19 The answer is **e. Estrogen** and **f. Progesterone.**

20 The answer is **d. GnRH.**

21 The answer is **a. FSH** and **b. LH.**

22 The answer is **b. LH.**

23 The answer is **e. Estrogen** and **f. Progesterone.**

24 The answer is **c. hCG.**

25 The answer is **a. Blastula.**

26 The answer is **d. Embryo.**

27 The answer is **e. Zygote.**

28 The answer is **g. Neonate.**

29 The answer is **c. Morula.**

30 The answer is **b. Fetus.**

31 The answer is **f. Blastocyst.**

32 The answer is **d. Gastrulation.**

33 The answer is **b. Morphogenesis.**

34 The answer is **c. Cleavage.**

35 The answer is **a. Differentiation of organ systems.**

Part V
Going Green with Plant Biology

DISCOVERY OF THE SQUIRTING FLOWER

"It appears to have all the physical properties of a normal plant... Hello—what's this?"

In this part . . .

Plants are everywhere, forming a background of green that people sometimes take for granted. But plants are essential to humans for many reasons: They make food, they provide oxygen, and people use them as a source of material for clothing and shelter. Beyond what people need, plants are amazing in their own right. Many people appreciate their beauty, but not everyone takes the time to look closely at the structure and function of plants. If they did, they'd be amazed at all the similarities they'd find between plants and people.

In this part, I lead you on a journey into the inner workings and structure of plants. You gain a deeper understanding of how they grow, live, and reproduce, and you get to test your new knowledge while you're at it.

Chapter 18

Studying Plant Structures

● ●

In This Chapter

▶ Understanding plant parts and their functions

▶ Breaking down the tissues of herbaceous and woody plants

▶ Following the steps of plant reproduction

● ●

A plant's structure suits its lifestyle. After all, it has flat leaves for gathering sunlight, roots for drawing water up from the soil, and flowers and fruits for reproduction. Plants begin their lives from seeds or spores, grow to maturity, and then reproduce asexually or sexually to create new generations. In this chapter, I present the fundamental structures of plants and introduce you to their reproductive strategies.

Peering at the Parts and Types of Plants

Like animals, plants are made of cells and tissues, and those tissues form organs, such as leaves and flowers, that are specialized for different functions. Plants have two basic organ systems:

 ✔ **The shoot system,** located above ground, helps plants capture energy from the sun for photosynthesis (see Chapter 4).

 ✔ **The root system,** located below ground, absorbs water and minerals from the soil.

The structure of each type of plant organ is tailored to match its function (see Figure 18-1):

 ✔ **Leaves** capture light and exchange gases with the atmosphere while minimizing water loss.

 • Many leaves are flattened so they have maximum surface area for light capture.

 • Tiny holes called *stomata* in the surfaces of leaves open and close to allow plants to absorb carbon dioxide from the atmosphere and to release oxygen. (You can see a stoma in the leaf cross section in Figure 18-1.)

 • *Guard cells* surround the stomata, ready to close them if water loss from the leaves becomes too great. The surface layer, or *epidermis,* of a leaf often has a coating of wax to further prevent water loss.

 ✔ **Stems** support leaves and reproductive structures and also transport sugars and waters throughout the plant.

 • Stems contain special types of tissues that give them strength. Woody plants have especially strong stems because they undergo *secondary growth* to thicken their stems and add layers of strong tissues.

- Stems contain tissues that specialize in transport. *Xylem* transports water from a plant's roots up to the leaves. *Phloem* transports sugars from the leaves throughout the plant. Young stems contain little packages of xylem and phloem, called *vascular bundles.*

✔ **Roots** grow through the soil, anchoring the plant and absorbing water and minerals.

- A *root cap* made of protective cells covers the tips of roots to prevent damage as they grow through the soil.

- The root's surface layer, also called an *epidermis,* contains cells that grow out into the soil, forming thin extensions called *root hairs.* These root hairs increase the root surface area so that the roots have more contact with the soil, which helps improve the absorption of water and minerals.

- Roots contain a core of vascular tissue that carries water away from the roots and toward the shoots and brings sugars from the shoots toward the roots. Some roots, like those of a carrot, specialize in storing extra sugars for later use by the plant.

✔ **Reproductive structures like flowers and cones** produce egg and sperm and may create protective structures around the young embryo. Flower structure (see Figure 18-2) also helps with *pollination,* the distribution of pollen (which contains sperm) to the plant's female parts.

- *Stamens* are the male parts of flowers. They consist of the *anther,* which makes pollen, and a thin stalk called a *filament.* Scientists call the ring of male parts within the flower the *androecium.*

- The flower's female parts are the *carpels,* which may be joined together to form a *pistil.* The part of the carpel that catches pollen is the *stigma,* and the swollen base that contains eggs in *ovules* is the *ovary.* Many flowers have an elongated tube, the *style,* between the stigma and the ovary. Scientists call the ring of female parts within the flower the *gynoecium.*

- The pretty parts of flowers are often showy *petals,* which help attract animals to flowers so they can help distribute pollen. Scientists call the ring of petals in the flower the *corolla.*

- Flowers may also have a ring of green, leaf-like structures called *sepals.* Sepals help protect the flower when it's still in the bud. In some flowers, the sepals look just like the petals and help attract pollinators. Scientists call the ring of sepals in the flower the *calyx.*

- After *fertilization* of the eggs by sperm, the ovules within a flower become seeds, and the ovary becomes a fruit. *Seeds* protect the embryo, and *fruits* help scatter the seeds away from the parent plant.

- A stalk called the *peduncle* supports the flower, which may also have a swollen base called the *receptacle.*

Based on the types of tissues they have and the reproductive structures they make, plants can be organized into four major groups:

✔ **Bryophytes** are plants such as mosses that don't have a vascular system and don't produce flowers or seeds.

✔ **Ferns and related plants** have vascular tissue, but they don't produce seeds.

✔ **Gymnosperms (also known as conifers)** have vascular tissue and produce cones and seeds, but they don't produce flowers.

✔ **Angiosperms (or flowering plants)** have vascular tissue and produce both flowers and seeds. Scientists divide the most familiar flowering plants into two groups based on the number of cotyledons they contain in their seeds. *Cotyledons,* sometimes called *seed leaves,* supply nutrition to the embryo and then emerge as the first leaves begin to grow.

- *Monocots,* like corn and lilies, have seeds that contain one cotyledon.

- *Dicots,* like beans, oak trees, and daisies, have seeds that contain two cotyledons.

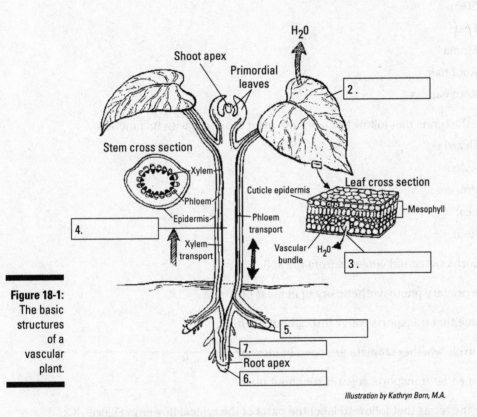

Figure 18-1:
The basic structures of a vascular plant.

Illustration by Kathryn Born, M.A.

In addition to their difference in seed structure, monocots and dicots have distinct patterns in their structures and the way they grow.

Table 18-1 presents several of the key structural differences between monocots and dicots.

Table 18-1	Structural Differences between Monocots and Dicots	
Feature	**Monocots**	**Dicots**
Cotyledons in seeds	One	Two
Bundles of vascular tissue in stem	Scattered throughout	Form definite ring pattern
Root system	Fibrous	Tap root
Leaf veins	Run parallel	Form a net pattern
Flower parts	Are in threes and multiples of threes	Are in fours and fives and multiples of fours and fives

1. Use two different colored pencils or highlighters to mark the shoot and root systems in Figure 18-1. Use one color to highlight the shoot system and the other color to highlight the root system.

2.–7. Use the terms that follow to label some of the main parts of the vascular plant in Figure 18-1.

a. Root

b. Stem

c. Leaf

d. Stoma

e. Root hair

f. Root cap

8.–12. Use the terms that follow to match the plant structure with its function.

a. Guard cell

b. Xylem

c. Phloem

d. Leaf

e. Root

8. Absorbs water and minerals from the soil.

9. The primary photosynthetic organ in most plants.

10. Tissue that transports water throughout plants.

11. Controls whether stomata are open or closed.

12. Tissue that transports sugar throughout plants.

13.–24. Use the terms that follow to label the parts of the typical flower in Figure 18-2.

a. Sepal

b. Filament

c. Anther

d. Stigma

e. Carpel (pistil)

f. Ovary

g. Peduncle

h. Petal

i. Receptacle

j. Ovule

k. Style

l. Stamen

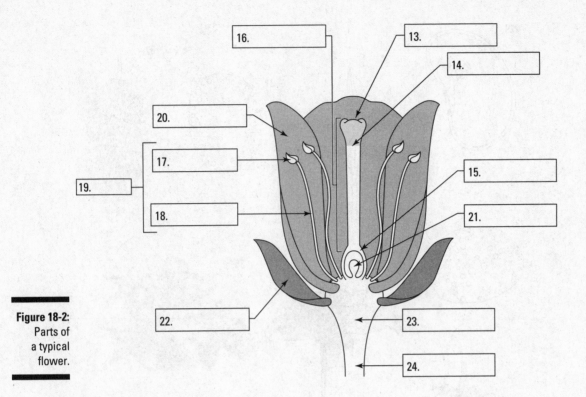

Figure 18-2:
Parts of
a typical
flower.

25.–28. Match the plant group to its description.

    a. Bryophytes

    b. Ferns and their allies

    c. Gymnosperms

    d. Angiosperms

  25. Plants that reproduce by cones and seeds.

  26. Plants that reproduce by flowers and seeds.

  27. Plants with neither vascular tissue nor seeds.

  28. Plants that have vascular tissue but don't make seeds.

29.–38. Use the terms that follow to identify the plant parts in Figure 18-3 and to indicate whether
the part belongs to a monocot or a dicot. Each structure has two correct answers.

    a. Monocot

    b. Dicot

    c. Stem

    d. Leaf

    e. Root

    f. Flower

    g. Seed

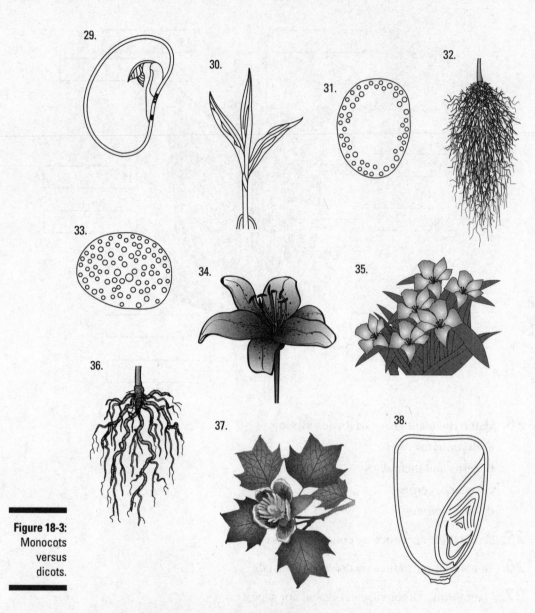

Figure 18-3:
Monocots
versus
dicots.

Taking a Look at Plant Tissues

Plant organs are made of plant tissues, which are made of plant cells (for the scoop on plant cells, see Chapter 3). All plants have tissues, but not all plants possess all three of the following types of tissues:

✔ **Dermal tissue:** Consisting primarily of epidermal cells, dermal tissue covers the entire surface of a plant.

✔ **Ground tissue:** This tissue type makes up most of a plant's body and contains three types of cells:

- *Parenchyma cells* are the most common ground tissue cells. They perform many basic plant cell functions, including storage, photosynthesis, and secretion.

- *Collenchyma cells* thicken their cell walls with extra cellulose to help support the plant.

- *Sclerenchyma cells* are similar to collenchyma cells, but their walls are even thicker and reinforced with *lignin,* a tough molecule found in wood. The cell walls of sclerenchyma cells are so thick, in fact, that mature sclerenchyma cells die because they can't get food or water across their walls via osmosis (more about osmosis in Chapter 3).

✔ **Vascular tissue:** You can think of vascular tissue as the plant's plumbing. The cells within xylem and phloem link up with one another end-to-end to form long columns of cells that carry nutrients and water up and down the plant.

- Xylem contains specialized cells called *vessels* and *tracheids.* These cells die at maturity, but their cell walls remain intact so that water can continue to flow. Vessel cells are wide and barrel-shaped, while tracheids are slimmer and have pointed ends.

- Phloem contains *sieve cells* for transporting sugars. Sieve cells remain alive but lose their nuclei at maturity as they become specialized for sugar transport. Nearby *companion cells* retain their nuclei and support the function of the sieve cells.

- Vascular tissue also contains parenchyma cells in the *vascular cambium,* a tissue of cells that can divide to produce new cells for the xylem and phloem.

Biologists use the appearance and feel of a plant's stem to place it into one of two categories: *herbaceous* (the stem remains somewhat soft and flexible) and *woody* (the stem has developed wood). All plant cells have *primary cell walls* made of cellulose, but the cells of woody plants have extra reinforcement from a *secondary cell wall* that contains lignin.

Plants that survive just one or two growing seasons — that is, *annuals* or *biennials* — are typically herbaceous plants. Plants that live year after year, called *perennials,* may become woody.

The stems of herbaceous and woody *dicots* (plants whose seeds contain two cotyledons; see the preceding section) are organized differently. You can see these differences most clearly if you look at a *cross section* (a section cut at right angles to the long axis) of a stem. (Imagine taking a hot dog and slicing it into little circles and you have a pretty good picture of how biologists make stem cross sections.)

When you look at a cross section of the stem of an herbaceous dicot, like the one in Figure 18-1, you see that

✔ The stem's center consists of *pith* (a soft, spongy tissue), which has many thin-walled cells called *parenchyma cells.* The thin walls allow the diffusion of nutrients and water among the cells.

✔ The vascular tissue is organized in vascular bundles that contain both xylem and phloem, as well as some vascular cambium. The vascular bundles are arranged in a ring around the pith.

✔ Outside the vascular bundle ring is the stem's *cortex.* It contains a layer of endodermis, additional parenchyma cells, and supporting tissue like collenchyma cells to help support the plant's weight and hold its stem upright.

✔ On the stem's surface are the epidermis and the cuticle, which is often waxed.

Woody dicots start life with green herbaceous stems that have vascular bundles. As they grow, however, the bundles merge with one another to form rings of vascular tissue that circle the stem. If you were to examine a cross section of the stem of a woody dicot that was a couple of years old, like the one in Figure 18-4, you'd see that

- The very center of the stem consists of a circle of pith.

- The xylem tissue forms a ring around the pith.

  - As woody plants grow, they add new layers of xylem every year, forming rings inside the woody stem. As these rings of xylem accumulate year after year, the woody stem's diameter increases.

  - During the spring, when lots of water is available, xylem vessels are larger, whereas during the drier summers, xylem vessels are smaller. The alternation of larger and smaller vessels gives wood a ringed appearance. You can count these rings in a tree's stem to tell how old it was when it was cut.

- Just outside the xylem rings is a thin ring of vascular cambium that's only one cell thick. As the stem grows, the vascular cambium divides to produce new xylem cells toward the inside of the stem and new phloem cells toward the outside of the stem.

- Outside the vascular cambium ring is a ring of phloem. The phloem of woody plants gets pushed farther and farther outward as the xylem tissue increases in size year after year. Phloem cells are fairly delicate, and the old phloem cells get crushed against the bark as the stem grows. The only phloem that serves to transport materials through the woody plant is the phloem that's newly formed during the most recent growing season. Phloem tissue is surrounded by strong cells called *fibers,* which are a type of sclerenchyma, and parenchyma cells that form the cortex.

- Outside the phloem ring is the *bark,* a ring of boxy, waterproof cells that help protect the stem. Bark includes the stem's outermost cells and a layer of cork cells just beneath that outermost layer. The *cork cambium* is a layer of parenchyma cells that divides to produce new cork cells, increasing the woody stem's diameter.

39.–43. Use the terms that follow to identify which type of tissue would perform the function in each question.

　　a. Parenchyma

　　b. Sclerenchyma

　　c. Collenchyma

　　d. Xylem

　　e. Phloem

39. A leaf cell does photosynthesis.

40. The cells that make the strings in celery thicken their cell walls with extra cellulose.

41. Sieve cells connect end to end to transport sugary sap through a tree trunk.

42. The cells that make the gritty texture in pears thicken their cell walls with lignin.

43. Hollow, open-ended cells called *vessels* conduct water through a flower stem.

44.–54. Use the terms that follow to label the woody dicot stem in Figure 18-4.

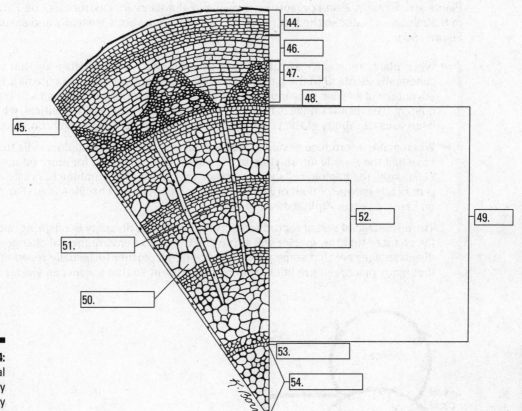

Figure 18-4:
Internal
anatomy
of a woody
stem.

Illustration by Kathryn Born, M.A.

    a. Pith

    b. Annual ring

    c. Phloem

    d. Cortex

    e. Cork cambium

    f. Vascular cambium

    g. Secondary xylem

    h. Primary xylem

    i. Summer wood

    j. Spring wood

    k. Cork

Growing Like a Weed: Plant Reproduction

Plants are very successful organisms, growing in almost every environment on Earth. Part of their success is due to the fact that they can reproduce both asexually and sexually (see Figure 18-5).

✔ When plants reproduce asexually, they use *mitosis* to produce offspring that are genetically identical to the parent plant (see Chapter 5 for more on mitosis). The advantage of asexual reproduction is that it allows successful organisms to reproduce quickly. The disadvantage is that all the offspring are genetically identical, which decreases the ability of the population to survive changes in the environment.

✔ When plants reproduce sexually, they use *meiosis* to produce haploid cells that have half the genetic information of the parent (see Chapter 5 for more on meiosis). Eventually, the haploid cells produce eggs and sperm that combine to create a new, genetically unique diploid organism that has two of every chromosome. (For the scoop on haploid versus diploid, head to Chapter 5.)

The advantage of sexual reproduction is that it creates diversity in offspring, increasing the chances that the species will survive in the face of environmental change. The disadvantages are that some plants need to find a partner to sexually reproduce and that many plants require liquid water to be present so that sperm can swim.

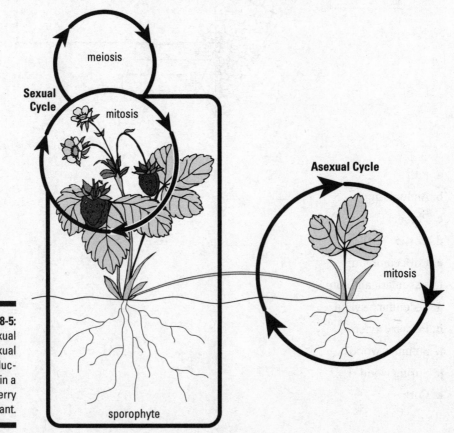

Figure 18-5: Asexual and sexual reproduction in a strawberry plant.

The life cycles of plants are a bit more complicated than those of animals. In animals, haploid cells called *gametes* (sperm and eggs) are usually small and inconspicuous. In plants, however, haploid cells can literally have a life of their own.

Plant life cycles involve an *alternation of generations:* The complete cycle includes two different generations called *sporophytes* and *gametophytes* (see Figure 18-6). Here's a breakdown of the cycle:

1. **Meiosis in a parent plant, called a *sporophyte,* results in the production of spores that are *haploid,* meaning they have half the genetic information of the parent plant.**

2. **The spores begin to grow by mitosis, developing into multicellular haploid organisms called *gametophytes.***

   The gametophyte step of the plant life cycle is a fundamental difference between plants and animals. In animals, no development occurs until a sperm and an egg combine to produce a new organism. In plants, there's a little break between meiosis and the production of sperm and eggs. During that break, a separate little haploid plant grows.

3. **Gametophytes produce gametes by mitosis.**

   In animals, meiosis produces sperm and egg, but in plants, meiosis occurs to produce the gametophyte. The gametophyte is already haploid, so it produces sperm and egg by mitosis.

4. **The gametes merge, producing cells called *zygotes* that contain the same number of chromosomes as the parent plant — that is, the zygotes are *diploid.***

5. **Zygotes divide by mitosis and develop into sporophytes so the life cycle can begin again.**

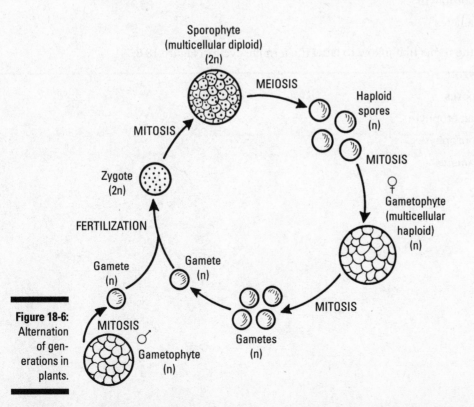

Figure 18-6: Alternation of generations in plants.

The plants you see when you go for a walk in the woods may be sporophytes or gametophytes:

✔ Mosses that grow on trees and on the forest floor are gametophytes. If you see little structures like flagpoles sticking off the moss, you're looking at a sporophyte. The little sporophytes grow like flags off the tops of the gametophytes. Inside the little flags, called *capsules,* meiosis is occurring to produce spores.

✔ Ferns you can see are sporophytes. If you look on the back of a fern's leaves, you can find little brown structures that seem dusty to the touch. These structures are where spores are being made, and the dust that comes off is the spores. Fern gametophytes are teeny — about as big as the fingernail on your pinkie — making them very tough to find in the wild.

✔ The conifers you see in a forest are sporophytes. The gametophyte generation in conifers is very small and contained within their cones.

✔ Flowering plants that are visible to the eye are also sporophytes. In flowering plants, the gametophyte generation is very small and contained within the flowers.

Practice your understanding of plant life cycles with the following questions.

55.–61. Use the terms that follow to label the moss life cycle in Figure 18-7.

    a. Zygote

    b. Spores

    c. Gametophyte

    d. Sporophyte

    e. Gametes

62.–66. Use the terms that follow to label the fern life cycle in Figure 18-8.

    a. Zygote

    b. Spores

    c. Gametophyte

    d. Sporophyte

    e. Gametes

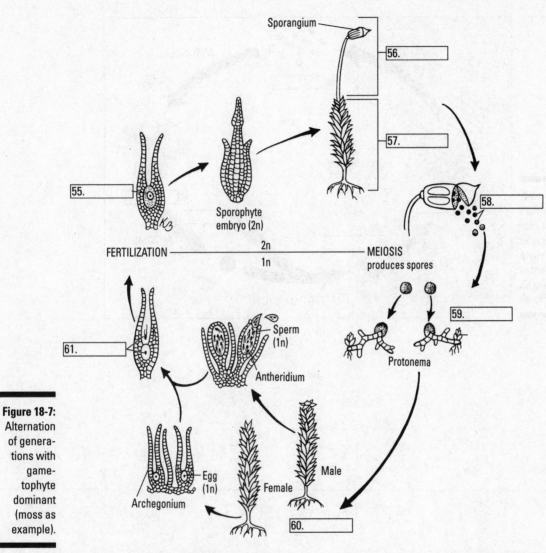

Sporangium

56.

57.

58.

55.

Sporophyte
embryo (2n)

FERTILIZATION —————— $\dfrac{2n}{1n}$ —————— MEIOSIS
produces spores

59.

Sperm
(1n)

61.

Antheridium

Protonema

Egg
(1n)

Archegonium

Female

Male

60.

Figure 18-7:
Alternation
of genera-
tions with
game-
tophyte
dominant
(moss as
example).

Illustration by Kathryn Born, M.A.

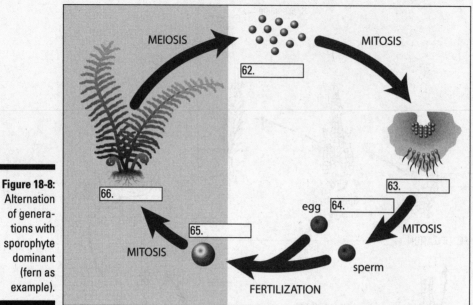

Figure 18-8:
Alternation of generations with sporophyte dominant (fern as example).

Answers to Questions on Plant Structures

The following are answers to the practice questions presented in this chapter.

1   You should have highlighted all the above-ground structures in one color and highlighted all the below-ground structures in a different color.

2–**7**   The following is how Figure 18-1 should be labeled:

2 **c. Leaf;** 3 **d. Stoma;** 4 **b. Stem;** 5 **e. Root hair;** 6 **f. Root cap;** 7 **a. Root.**

8   The answer is **e. Root.**

9   The answer is **d. Leaf.**

10   The answer is **b. Xylem.**

11   The answer is **a. Guard cell.**

12   The answer is **c. Phloem.**

13–**24**   The following is how Figure 18-2 should be labeled:

13 **d. Stigma;** 14 **k. Style;** 15 **f. Ovary;** 16 **e. Carpel (pistil);** 17 **c. Anther;** 18 **b. Filament;** 19 **l. Stamen;** 20 **h. Petal;** 21 **j. Ovule;** 22 **a. Sepal;** 23 **i. Receptacle;** 24 **g. Peduncle.**

25   The answer is **c. Gymnosperms.**

26   The answer is **d. Angiosperms.**

27   The answer is **a. Bryophytes.**

28   The answer is **b. Ferns and their allies.**

29–**38**   The following is how Figure 18-3 should be labeled:

29 **b. Dicot** and **g. Seed;** 30 **a. Monocot** and **d. Leaf;** 31 **b. Dicot** and **c. Stem;** 32 **a. Monocot** and **e. Root;** 33 **a. Monocot** and **c. Stem;** 34 **a. Monocot** and **f. Flower;** 35 **b. Dicot** and **f. Flower;** 36 **b. Dicot** and **e. Root;** 37 **b. Dicot** and **d. Leaf;** 38 **a. Monocot** and **g. Seed.**

39   The answer is **a. Parenchyma.**

40   The answer is **c. Collenchyma.**

41   The answer is **e. Phloem.**

42   The answer is **b. Sclerenchyma.**

43   The answer is **d. Xylem.**

44–54 The following is how Figure 18-4 should be labeled:

44 k. Cork; 45 e. Cork cambium; 46 d. Cortex; 47 c. Phloem; 48 f. Vascular cambium; 49 g. Secondary xylem; 50 j. Spring wood; 51 i. Summer wood; 52 b. Annual ring; 53 h. Primary xylem; 54 a. Pith.

55–61 The following is how Figure 18-7 should be labeled:

55 a. Zygote; 56 d. Sporophyte; 57 c. Gametophyte; 58 b. Spores; 59 c. Gametophyte; 60 c. Gametophyte; 61 e. Gametes.

62–66 The following is how Figure 18-8 should be labeled:

62 b. Spores; 63 c. Gametophyte; 64 e. Gametes; 65 a. Zygote; 66 d. Sporophyte.

Chapter 19

Pondering Problems in Plant Physiology

Plant growth has many things in common with animal growth. Just like you, plants need food, water, and minerals in order to grow. (Of course, a big difference between you and a plant is that plants make their own food by photosynthesis.) Just like you have a circulatory system that moves blood around your body, plants have tissues that move fluids around their bodies. And just like hormones can have a big effect on your growth and development, plant hormones trigger many aspects of plant growth that you're probably familiar with. In this chapter, I present the fundamentals of *plant physiology,* the study of how plants function. I explain the processes plants use to transport fluids around their bodies, and I introduce you to some of the major effects of plant hormones.

Taking Minerals from the Soil

Just like you do, plants build their cells from carbohydrates, proteins, lipids, and nucleic acids (see Chapter 3 for more on cells and molecules). The difference between you and a plant is that you get all these molecules from your food, but plants need to build them for themselves. Plants get all the carbon, hydrogen, and oxygen they need from the carbohydrates they build during photosynthesis (see Chapter 4), but to build other kinds of molecules they also need mineral elements like nitrogen, phosphorous, and sulfur. Plants get these and other minerals from the soil.

The mineral nutrients found in soil dissolve in water. When plants absorb water through their roots, they obtain both macronutrients and micronutrients. *Macronutrients* help with molecule construction, and *micronutrients* act as partners for enzymes and other proteins to help them function. Plants generally require large amounts of macronutrients and smaller amounts of micronutrients. Table 19-1 lists the specific macronutrients and micronutrients plants absorb from soil.

Table 19-1	The Essential Nutrients Plants Pull from Soil
Macronutrients	*Micronutrients*
Calcium (Ca)	Boron (B)
Magnesium (Mg)	Chloride (Cl)
Nitrogen (N)	Copper (Cu)
Phosphorous (P)	Iron (Fe)
Potassium (K)	Manganese (Mn)
Sulfur (S)	Molybdenum (Mb)
	Zinc (Zn)

You can remember the most important elements for plants with the phrase *C. Hopkins Café, Mighty Good.* The *CHOPKNS CaFe Mg* stands for carbon, hydrogen, oxygen, phosphorous, potassium, nitrogen, sulfur, calcium, iron, and magnesium. All these elements are macronutrients for plants, with the exception of iron, which is considered a micronutrient.

If plants don't get enough of one of these important elements, they can't function correctly. Without carbon, hydrogen, and oxygen (from carbon dioxide and water), plants can't grow at all. And even though plants need smaller amounts of minerals, each missing mineral causes a specific problem.

Test your understanding of plant nutrition with the following questions.

1.–5. Use the following terms to identify whether the nutrient is a macronutrient or a micronutrient.

    a. Macronutrient

    b. Micronutrient

1. A plant gets sulfur from sulfate (SO_4^-) in the soil.

2. A plant takes in calcium as calcium salts from the soil.

3. A plant takes in nitrogen as nitrates (NO_3^{2-}) from the soil.

4. A plant takes in magnesium (Mg) from the soil.

5. A plant absorbs copper (Cu) from the soil.

6. True or false: Plants get carbon from the soil.

7. True or false: A plant's weight comes mostly from the minerals it takes from the soil.

Pulling Water Through Plants

Several processes work together to transport water from where a plant absorbs it (the roots) upward through the rest of its body. To understand how these processes work, you first need to know one key feature of water: Water molecules tend to stick together, literally.

Water molecules are attracted to one another and to surfaces by weak electrical attractions. When water molecules stick together by hydrogen bonds (see Chapter 2), scientists call it *cohesion*. When water molecules stick to other materials, scientists call it *adhesion*.

A familiar example of the stickiness of water occurs when you drink water through a straw — a process that's very similar to the method plants use to pull water through their bodies. You apply suction at the top of the straw, and the water molecules move toward your mouth. Because the molecules cling to each other on the sides of the straw, they stay together in a continuous column and flow into your mouth.

Scientists call the explanation for how water moves through plants the *cohesion-tension theory*. It involves three main factors:

- **Transpiration:** *Transpiration* is the technical term for the evaporation of water from plants. As water evaporates through the stomata in the leaves (or any part of the plant exposed to air), it creates a negative pressure (also called *tension* or *suction*) in the leaves and tissues of the xylem. The negative pressure exerts a pulling force on the water in the plant's xylem and draws the water upward (just like you draw water upward when you suck on a straw).

- **Cohesion:** When water molecules stick to one another through *cohesion*, they fill the column in the xylem and act as a huge single molecule of water (like water in a straw).

- **Capillary action:** *Capillary action* is the movement of a liquid across the surface of a solid caused by adhesion between the two. When you a place a tube in water, water automatically moves up the sides of the tube because of adhesion, even before you apply any sucking force. The narrower the tube, the higher the water climbs on its own. In plants, adhesion forces water to be pulled up the columns of cells in the xylem and through fine tubes in the cell wall.

Environmental conditions like heat, wind, and dry air can increase the rate of transpiration from a plant's leaves, causing water to move more quickly through the xylem. Sometimes, the pull from the leaves is stronger than the weak electrical attractions among the water molecules, and the columns of water can break, causing air bubbles to form in the xylem.

The sudden appearance of gas bubbles in a liquid is called *cavitation*.

To repair the lines of water, plants create *root pressure* to push water up into the xylem. At night, root cells release ions into the xylem, increasing its solute concentration. Water flows into the xylem by osmosis (see Chapter 3), pushing a broken water column up through the gap until it reaches the rest of the column.

If environmental conditions cause rapid water loss, plants can protect themselves by closing their stomata. However, after the stomata are closed, plants don't have access to carbon dioxide (CO_2) from the atmosphere, which shuts down photosynthesis (see Chapter 4). Some plants, like those that live in deserts, must routinely juggle between the competing demands of getting CO_2 and not losing too much water.

8.–12. Use the terms that follow to demonstrate the movement of water through plants by labeling Figure 19-1.

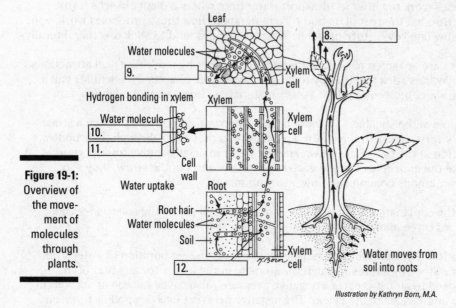

Figure 19-1:
Overview of the movement of molecules through plants.

Leaf

Water molecules

8.

9.

Xylem cell

Hydrogen bonding in xylem

Xylem

Water molecule

Xylem cell

10.

11.

Cell wall

Water uptake

Root

Root hair

Water molecules

Soil

Xylem cell

12.

Water moves from soil into roots

Illustration by Kathryn Born, M.A.

    a. Cohesion

    b. Osmosis

    c. Adhesion

    d. Transpiration

    e. Stomata

13. Some desert plants protect themselves from water loss by doing a special kind of photosynthesis called *crassulacean acid metabolism* (CAM) *photosynthesis.* CAM plants open their stomata and gather carbon dioxide at night. They attach the CO_2 to other molecules for storage and then release it again during the day, when light is available for photosynthesis. How does the strategy of CAM photosynthesis protect the plants from water loss?

Sending Sugars from Sources to Sinks

Phloem, similar to blood vessels, transports *sap* — a sticky solution that contains sugars, water, minerals, amino acids, and plant hormones — throughout the plant via *translocation,* the transport of dissolved materials in a plant. Unlike the xylem, which can only carry water upward, phloem carries sap upward and downward, from sugar sources to sugar sinks:

✔ **Sugar sources** are plant organs such as leaves that produce sugars.

✔ **Sugar sinks** are plant organs such as roots, tubers (underground stems), and bulbs (swollen leaves) that consume or store sugars.

Scientists call their explanation for how translocation works in a plant's phloem the *pressure-flow hypothesis.* Figure 19-2 illustrates this hypothesis, the steps of which I explain here:

1. **Sugars, produced within sugar sources, are loaded into phloem cells called *sieve tube elements,* creating a high concentration.**

2. **Water enters the sieve tube elements by osmosis.**

   During osmosis, water moves into the areas with the highest concentration of solutes (in this case, sugars).

3. **The inflow of water increases pressure at the source, causing the movement of water and carbohydrates toward the sieve tube elements at a sugar sink.**

   You can think of this step like turning on a water faucet that's connected to a garden hose. As water flows from the tank into the hose, it pushes the water in front of it down the hose.

4. **Sugars are removed from cells at the sugar sink, keeping the concentration of sugars low.**

   As a sugar sink receives water and carbohydrates, pressure builds. But before the sugar sink can turn into a sugar source, carbohydrates in the sink are actively transported out of the sink and into needy plant cells. As the carbohydrates are removed, the water then follows the solutes and diffuses out of the cell, relieving the pressure.

Sugar sinks that store carbohydrates can become sugar sources for plants when the plants need sugar. *Starch,* a complex carbohydrate, is insoluble in water, so it acts as a carbohydrate storage molecule. Whenever a plant needs sugar, like at night or in the winter, when photosynthesis doesn't occur as well, the plant can break down its starches into simple sugars, which allows a tissue that would normally be a sugar sink to become a sugar source.

Because plant cells can act as both sinks and sources, and because phloem transport goes both upward and downward, plants are pretty good at spreading the wealth of carbohydrates and fluid to where the plants need them. As long as a plant has a continuous incoming source of minerals, water, carbon dioxide, and light, it can fend for itself.

Practice your understanding of sugar movement in plants by answering the following questions.

14.–16. Use the following terms to describe the scenario in each question. Each question has two correct answers.

   a. Sink for sugar

   b. Source for sugar

   c. Low *turgor pressure* (pressure against the cell wall)

   d. High turgor pressure

14. A rapidly growing plant bud

15. A healthy plant leaf on a sunny day

16. A growing plant root

17. A young plant embryo is just beginning to grow out of its seed. It's still underground, so no light can reach it. The embryo is surrounded by starch that was placed in the seed by the parent plant. Use the pressure-flow hypothesis to explain how sugar can move from the seed into the cells of the plant embryo.

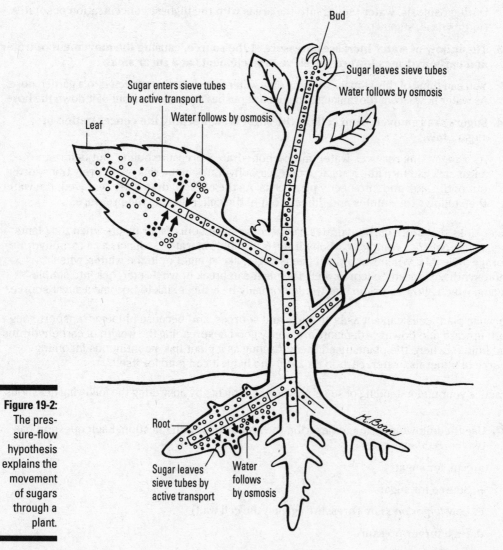

Figure 19-2:
The pressure-flow hypothesis explains the movement of sugars through a plant.

Illustration by Kathryn Born, M.A.

Sending Signals with Plant Hormones

Plant cells communicate with one another via messengers called *hormones,* chemical signals produced by one type of cell that travel to target cells and cause changes in their growth or development. Plant hormones control many familiar plant behaviors.

Six categories of hormones control plant growth and development:

- **Auxins** stimulate the elongation of cells in the plant stem and *phototropism* (the growth of plants toward light). If a plant receives equal light on all sides, its stem grows straight. If light is uneven, auxin moves toward the plant's darker side and causes cells on that side to lengthen. This may seem backward, but when the shady side of the stem grows, the stem, in its crookedness, actually bends toward the light. This action keeps the leaves positioned toward the light so photosynthesis can continue.

  Auxin also inhibits buds on the sides of plants, called *lateral buds,* from growing into branches. Auxin is produced by the *apical meristem,* a region of dividing cells at the tip of the main branch. Because it inhibits lateral buds, auxin establishes *apical dominance;* that is, growth of the main shoot is favored over growth of the side shoots. As the main shoot's tip gets farther away from side shoots, they begin to grow.

- **Gibberellins** promote both cell division and cell elongation, causing shoots to elongate so that plants can grow taller and leaves can grow bigger. They also signal buds and seeds to begin growing in the spring, and they promote flowering.

- **Cytokinins** stimulate cell division, promote leaf expansion, and slow down the aging of leaves. Florists actually use them to help make cut flowers last longer.

- **Abscisic acid** inhibits cell growth and can help prevent water loss by triggering stomata to close. Plant nurseries use abscisic acid to keep plants dormant during shipping.

- **Ethylene** stimulates the ripening of fruit and signals deciduous trees to drop their leaves in the fall. Fruit growers use ethylene to partially ripen fruit for sale.

- **Brassinosteroids** control many aspects of plant growth and development. They're extremely powerful and can affect the concentrations of other plant hormones, promoting or inhibiting plant growth depending on the stage of development. They can counteract the effects of abscisic acid, acting to promote flowering, seed germination, and the opening of stomata. Scientists are very interested in brassinosteroids because they have the ability to increase photosynthesis, an ability that may come in handy in agriculture!

Some of the flavor-making processes that occur in fruits happen while the fruits are still on the plant. So even though ethylene can trigger some parts of ripening, like softening a fruit after it has been picked, fruit that's picked unripe doesn't taste as good as fruit that has ripened on the plant. That's why you can buy a big, beautiful tomato at the grocery store and take it home, only to discover that it doesn't have much flavor. It was probably picked unripe and then treated with ethylene.

18.–24. Use the following terms to indicate which plant hormone is triggering the growth or behavior of the plant.

  a. Auxin

  b. Cytokinin

  c. Gibberellin

  d. Abscisic acid

  e. Ethylene

  f. Brassinosteroids

18. A seed germinates in the spring.

19. It's autumn, and apples are ripening on the trees.

20. It's autumn, and leaves are turning red and gold.

21. It's summer, and a bud opens to reveal a beautiful rose.

22. It's spring, and bright green leaves begin growing out of the buds on trees.

23. A houseplant on your windowsill leans its stems toward the window.

24. You pinch off the tip of one of your houseplants to encourage the plant to grow bushier instead of taller.

Answers to Questions on Plant Physiology

The following are answers to the practice questions presented in this chapter.

1 The answer is **a. Macronutrient.**

2 The answer is **a. Macronutrient.**

3 The answer is **a. Macronutrient.**

4 The answer is **a. Macronutrient.**

5 The answer is **b. Micronutrient.**

6 The answer is **false.**

Plants get carbon from the air as carbon dioxide.

7 The answer is **false.**

Although plants take minerals from the soil, the amount of these minerals is very small compared to the proteins, carbohydrates, lipids, and nucleic acids that make up the plant's body. All these big molecules have a carbon backbone (see Chapter 2), so carbon atoms make up the majority of a plant's weight. Plants get carbon from carbon dioxide in the atmosphere. (If you said true to this question, you're not alone; many people share this misunderstanding about where plants get most of the material they need to grow. Flip to Chapter 4 and check out the details on photosynthesis to discover how plants make the food molecules they use to build their bodies.)

8 – 12 The following is how Figure 19-1 should be labeled:

8 **d. Transpiration;** 9 **e. Stomata;** 10 **c. Adhesion;** 11 **a. Cohesion;** 12 **b. Osmosis.**

13 Temperatures are typically cooler at night, so water doesn't evaporate as quickly, which lowers the rate of transpiration. When CAM plants open their stomata at night, they can collect carbon dioxide with minimal water loss. Then, during the day when sunlight is available for photosynthesis but temperatures go up, CAM plants can release the CO_2 inside their leaves and do photosynthesis while keeping their stomata safely shut!

14 The answer is **a. Sink for sugar** and **c. Low turgor pressure.**

15 The answer is **b. Source for sugar** and **d. High turgor pressure.**

A healthy plant leaf on a sunny day does lots of photosynthesis.

16 The answer is **a. Sink for sugar** and **c. Low turgor pressure.**

A root growing through the soil can't do photosynthesis.

17   The embryo needs sugar to grow, but it can't make it for itself. So it's a sink for sugar and has low turgor pressure. Enzymes break down the starch stored in the seed, converting it to sugars. This makes the seed a source for sugars. The high concentration of sugar in the seed attracts water through osmosis, resulting in high turgor pressure. The pressure forces fluid carrying sugar to move from the seed to the embryo. The embryo's cells snatch up the sugars and use them to grow.

18   The answer is **c. Gibberellin.**

Gibberellin promotes seed germination. Germination can also be enhanced by **f. Brassinosteroids.**

19   The answer is **e. Ethylene.**

Ethylene triggers the ripening of fruit like apples.

20   The answer is **e. Ethylene.**

Ethylene triggers the senescence (aging) of leaves.

21   The answer is **c. Gibberellin.**

Gibberellin promotes flowering. Flowering can also be enhanced by **f. Brassinosteroids.**

22   The answer is **b. Cytokinin.**

Cytokinin promotes leaf expansion.

23   The answer is **a. Auxin.**

Auxin triggers phototropism, causing plants to bend toward the light.

24   The answer is **a. Auxin.**

Auxin establishes apical dominance. By pinching off the main shoot's tip, you stop the production of auxin and encourage the side shoots to grow.

Part VI
The Part of Tens

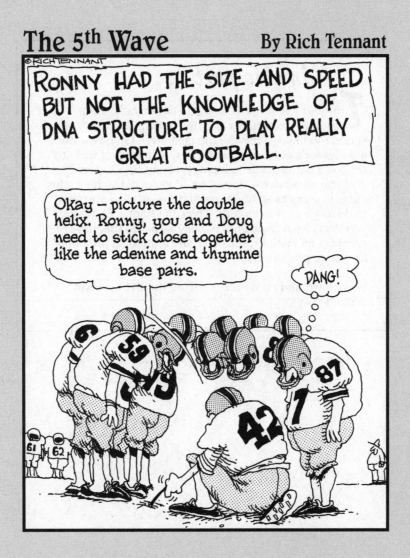

In this part . . .

This part has some extra-light (but extra-useful!) reading — resources and tips to help you take your success in biology class to the next level. First, how does a professor's personal cheat sheet, chock-full of ways to get you a better grade in biology, sound? In Chapter 20, I give you my advice — based on 14 years of teaching, plus my own years as an undergraduate — on what you need to do to get an A in biology. Then, in Chapter 21, I point you toward ten (plus one) biology websites that have excellent animations, tutorials, and additional problems for you to practice on.

Work hard and enjoy discovering the living world around you!

Chapter 20

Ten Tips for Getting an A in Biology

In This Chapter
▶ Studying frequently and actively
▶ Preparing yourself for tests
▶ Asking for help and using your resources

Science classes may be among the most challenging classes you'll ever take. I've had many conversations with students over the years and read research on student learning, and I think I've figured out some of the major issues you'll face. In this chapter, I present ten tips to help you deal with those challenges and get the best grade you can.

Put in Your Time

One of the reasons that science classes are so challenging is that they ask you to look at things you've never looked at before. When you take an English class or even a psychology class, you're often on familiar ground, adding to knowledge that you've gained from former classes or other sources.

In science classes, however, you may be starting almost from scratch. So the first tip I have for you is this: Plan to spend enough time studying, much more time than you may be used to for a different kind of class.

The rule of thumb for a science class is to budget two hours outside of class for every hour that you're in class. For a class that meets five hours a week, that's ten extra hours of study time — *just* for that class. To find out what you should be doing during all that study time, keep reading.

Make Vocabulary Flashcards

Studies have shown that you learn more new words in a first-year biology class than you do in a first-year language class. That's a lot of terminology. And believe me, instructors will introduce a term to you just once and then test you on it later. So if you rely only on your classroom experience to reinforce the new terms that you need to know, you'll come up short on the first exam. And that's a shame, because for most people, memorizing is the easiest level of learning. The bottom line is that if you don't memorize all the new terms, you'll miss out on some easy points that can help buffer your exam grade.

Flashcards are a great memorization aid. Spend at least one hour of your study time a week making and studying flashcards. Put the new term on one side of the card and the definition on the other side.

Be creative in how you define the words. The definitions in your textbook's glossary may be just as foreign to you as the terms themselves, so write the definitions in your own words. Include a clue for yourself on how to remember the word — perhaps a drawing if the term describes a new structure, or just anything to jog your memory.

When you study, go through your stack of cards both ways. Read the terms and say the definitions on one pass, and read the definitions and give the terms on the next pass.

Pace Yourself

Your brain has two kinds of memory — short-term and long-term. Have you ever listened to your instructor explain something and thought, "Cool, I totally get that," but then later found yourself scratching your head trying to remember the details of what you learned? That's because you had the idea or process in your short-term memory but didn't get it fixed in your long-term memory. When you sleep, your brain weeds through the information you collected that day and dumps a lot of the stuff that was in your short-term memory. (If our brains didn't do this, we'd all be on information overload.)

You have to outsmart your brain. Budget a small amount of study time every day instead of planning on big marathon sessions once a week. If you review your lecture notes on the day that you first wrote them down, while the info is still fresh in your short-term memory, the notes will be clearer, and you'll increase your chances of banking some of that information in your long-term memory before you go to sleep.

Study Actively, Not Passively

Reading is good, and reading your textbook definitely helps improve your understanding of the material. But reading alone won't get most people a good grade in a science class. To store information in your long-term memory, you have to use the information actively.

For example, think of something in your life that you're good at. Do you play an instrument? Are you a good cook? Are you an athlete? Whatever it is you're good at, I'm willing to bet you gained your skills by doing, not just reading. You need to take the same approach with your biology class if you want to get a good grade.

You can practice what you learn in several ways:

- ✔ **Do the activities in lab.** Hands-on laboratory experiments help reinforce concepts from class— if, that is, you actually show up to lab, do the experiments, and ask questions.
- ✔ **Draw processes and structures.** Take out some blank paper and try to draw the things you're learning about. Label everything and explain the concepts to yourself as you go along. Peek at your notes when you have to, but keep repeating the process until you don't have to peek anymore.

✔ **Explain things to others.** If you study alone, you can explain things out loud to yourself, your significant other, your parents, your kids, or even your cat.

✔ **Answer questions at the back of your book chapter.** Instructors often recommend questions to go along with the reading. These questions are good practice, especially the critical thinking questions.

Phone a Friend

Study groups can really improve your success in science classes. You can practice your explanations on people who are studying the same material, ask and answer questions, and share tips and tricks with one another. You can also support one another emotionally and — gasp! — maybe even make studying more fun. Many of my students form study groups that stay together through a whole year of classes, and sometimes even longer.

Have fun, but stay focused! If chatting about nonscience subjects takes up too much of your group's time, then either refocus the group, find a new group, or stop inviting whoever's disrupting the group.

Test Yourself Before Your Instructor Tests You

After you take a test, your instructor records your performance permanently in her grade book, influencing your final grade in the course. You don't want the exam to be your first clue about whether you really know and understand the information! Before you take the test, find ways to test yourself and to identify your weak spots so you can make sure you're really ready.

Here are some tips on how to test yourself:

✔ **Some instructors actually give copies of old exams to students to practice on.** Ask your instructor if she does this.

✔ **Textbooks have quizzes at the back of the chapters and often have online companion sites with more quizzes.**

But be warned — your instructor didn't write these quizzes, and they can sometimes be too easy. If the quizzes that come with your book all seem like factual questions (memorized information), they may give you a false sense of security. That doesn't mean they're useless — they can check how well you know the facts — but if your instructor also values critical thinking, you have to find other sources as well.

Maximize the Easy Points

Getting a good grade is about getting the best overall percentage in class that you can. Exam points are usually the hardest to get, so make sure you get all the easy points, which usually come from homework assignments, labs, attendance, and even extra credit. Take advantage of every easy assignment that comes your way. Then, if you miss a few exam points, you've got backup.

Ask for Help Upfront

Semesters go by fast, and quarters go by even faster. Don't wait until it's too late to get help. At the first sign of trouble, like a bad grade on an assignment or quiz, get help from your instructor, your teaching assistant, the tutoring center, or a friend who's doing well in the class.

Don't be afraid to ask questions or ask for help. You may be embarrassed to tell someone you don't understand, but try not to let those feelings keep you from getting help when you need it. You're not supposed to be an expert on the subject; that's why you're taking the class. Sometimes, the question you ask is the one that ten other people are wondering about. Instructors and TAs are paid to help you learn, and most of them love what they do.

Use Your Resources

Traditional biology textbooks cost big bucks, but they also come with some nifty add-ons like access to websites with animations, quizzes, and tutorials. Sometimes, a good animation is worth a thousand words, so check out your resources and incorporate the good ones into your study routine. If your book doesn't come with these bells and whistles, you can still find lots of good material on the Internet. YouTube is loaded with animations and even student-created songs to help you memorize something.

Be careful with materials that aren't created by a publisher or a scientist. I've spotted errors in some homemade materials.

Don't Leave It in the Classroom

Research on human learning shows that people remember information best when they understand its importance — in other words, when the info is connected to a fundamental concept that's part of their existing knowledge. Traditional science classes don't always do a good job of helping students make these necessary connections. Information can be thrown at you rapidly without you having a chance to let it sink in and connect it to what you know.

The whole point of science is to help people understand their world better. So don't leave what you learn in the classroom!

Chapter 21

Ten (Plus One) Great Biology Websites

In This Chapter

▶ Finding online resources to help you understand difficult concepts

▶ Testing yourself using worksheets, practice problems, and quizzes

Sometimes, your textbook (and even your instructor) just doesn't make sense. Luckily for you, you have a *For Dummies* book *and* the Internet at your fingertips. Several excellent websites contain tutorials and practice problems to help you master the subject of biology.

In this chapter I give you ten (or so) of my favorites, with a preference for those sites that include access to practice problems and quizzes.

Solving problems gives you practice applying what you learn in class and gives you a double benefit:

✔ Practicing the information helps you store it in your long-term memory. In other words, you learn the information better.

✔ Practice problems are great training for problems your instructor may throw at you on an exam. You don't want the exam to be the first time you try to do something!

I've tried to include only sites that seem relatively stable over the long term, but you can find many more problem sets from various biology courses if you use an Internet search engine.

Dummies.com

www.dummies.com/

You bought this book, so you obviously appreciate the *For Dummies* approach to making information straightforward. The *For Dummies* website offers you more help in the form of biology cheat sheets and many short articles on various topics from class that you need to understand.

Use the topics toolbar on the left side of the page to navigate to Education & Languages and then Science. From Science, you can access Biology, Botany, and Anatomy & Physiology materials.

The Biology Project at the University of Arizona

www.biology.arizona.edu/

This site has great tutorials and problems on a range of biology topics, mostly focused around cell and human biology. Many of the problem sets are interactive, so if you get a problem wrong, the site automatically sends you to the right tutorial to help you figure out why.

Genetic Science Learning Center

http://learn.genetics.utah.edu/

This site has outstanding tutorials and essays focused on the topics of DNA and genetics. As you explore the topics on this site, you encounter activities that let you test your understanding as you go along.

DNA from the Beginning

http://dnaftb.org/

Another terrific site focused on DNA and genetics. The site is organized around concepts. To support each concept, the site has a combination of easy-to-read content, animations, videos, and problems to solve.

Life: The Science of Biology

http://bcs.whfreeman.com/thelifewire/default.asp

This website accompanies a textbook published by W.H. Freeman, but it's accessible without a password. The tutorials on this site are some of the best I've seen on the web, and they include pre- and post-tests.

The Biology Corner

www.biologycorner.com/

This blog from a high school biology teacher has some excellent worksheets that you can print and practice on for free. The only downside is that she doesn't post the keys; her students have access to her site, too, so she doesn't want to give away the answers. These worksheets may be best for working in study groups so you can check your answers with those of other students.

Cells Alive!

www.cellsalive.com/

This site has some great videos of different kinds of cells, as well as some interactive animations on subjects like cell division and parts of the cell.

The Virtual Cell

www.ibiblio.org/virtualcell/

Click the Virtual Cell Tour to navigate within a virtual cell and receive prompts that explain the structures within. You can also download a worksheet to complete as you take the tour.

The Virtual Plant Cell

www.life.illinois.edu/plantbio/cell/

This website lets you zoom in on cellular structures by clicking the cell. You change from zoom to other actions like cutting the cell by changing the radio button to the right of the figure before you click. By choosing the EM Image radio button, you bring up actual photographs of cellular structures taken through an electron microscope.

Quia.com

www.quia.com/shared/biology/

This site has fun flashcards and games to test you on a variety of subjects. Some of these activities are available without a membership. If you do want a membership, the site offers a 30-day free trial.

HHMI BioInteractive Holiday Lectures

www.hhmi.org/biointeractive/hl/

The Howard Hughes Medical Institute sponsors a holiday lecture series at the end of every year, inviting speakers to travel to an undergraduate institution and present lectures on a chosen topic. The best part is that all these lectures are videotaped and posted online after the event so that any interested person can watch them for free!

Some of the topics chosen have been very interesting; the most recent was on the origin of humans. Some of my past favorites include lectures on AIDS ("AIDS: Evolution of an Epidemic") and infectious diseases ("Confronting the Microbe Menace"). Other lectures include "Making Your Mind: Molecules, Motion, and Memory" (on the human brain) and "The Science of Fat" (on the obesity epidemic).

These lectures are generally well presented by scientists who are also good speakers. Watching them is an excellent way to learn more about how science connects to your life. HHMI also has other types of resources available at its BioInteractive site. Just look at the menu bar along the top of the page and you'll see links to videos, animations, and virtual labs that really help biology come alive!

Index

Business/Accounting & Bookkeeping

Bookkeeping For Dummies
978-0-7645-9848-7

eBay Business
All-in-One For Dummies,
2nd Edition
978-0-470-38536-4

Job Interviews
For Dummies,
3rd Edition
978-0-470-17748-8

Resumes For Dummies,
5th Edition
978-0-470-08037-5

Stock Investing
For Dummies,
3rd Edition
978-0-470-40114-9

Successful Time
Management
For Dummies
978-0-470-29034-7

Computer Hardware

BlackBerry For Dummies,
3rd Edition
978-0-470-45762-7

Computers For Seniors
For Dummies
978-0-470-24055-7

iPhone For Dummies,
2nd Edition
978-0-470-42342-4

Laptops For Dummies,
3rd Edition
978-0-470-27759-1

Macs For Dummies,
10th Edition
978-0-470-27817-8

Cooking & Entertaining

Cooking Basics
For Dummies,
3rd Edition
978-0-7645-7206-7

Wine For Dummies,
4th Edition
978-0-470-04579-4

Diet & Nutrition

Dieting For Dummies,
2nd Edition
978-0-7645-4149-0

Nutrition For Dummies,
4th Edition
978-0-471-79868-2

Weight Training
For Dummies,
3rd Edition
978-0-471-76845-6

Digital Photography

Digital Photography
For Dummies,
6th Edition
978-0-470-25074-7

Photoshop Elements 7
For Dummies
978-0-470-39700-8

Gardening

Gardening Basics
For Dummies
978-0-470-03749-2

Organic Gardening
For Dummies,
2nd Edition
978-0-470-43067-5

Green/Sustainable

Green Building
& Remodeling
For Dummies
978-0-470-17559-0

Green Cleaning
For Dummies
978-0-470-39106-8

Green IT For Dummies
978-0-470-38688-0

Health

Diabetes For Dummies,
3rd Edition
978-0-470-27086-8

Food Allergies
For Dummies
978-0-470-09584-3

Living Gluten-Free
For Dummies
978-0-471-77383-2

Hobbies/General

Chess For Dummies,
2nd Edition
978-0-7645-8404-6

Drawing For Dummies
978-0-7645-5476-6

Knitting For Dummies,
2nd Edition
978-0-470-28747-7

Organizing For Dummies
978-0-7645-5300-4

SuDoku For Dummies
978-0-470-01892-7

Home Improvement

Energy Efficient Homes
For Dummies
978-0-470-37602-7

Home Theater
For Dummies,
3rd Edition
978-0-470-41189-6

Living the Country Lifestyle
All-in-One For Dummies
978-0-470-43061-3

Solar Power Your Home
For Dummies
978-0-470-17569-9

Internet
Blogging For Dummies,
2nd Edition
978-0-470-23017-6

eBay For Dummies,
6th Edition
978-0-470-49741-8

Facebook For Dummies
978-0-470-26273-3

Google Blogger
For Dummies
978-0-470-40742-4

Web Marketing
For Dummies,
2nd Edition
978-0-470-37181-7

WordPress For Dummies,
2nd Edition
978-0-470-40296-2

Language & Foreign Language
French For Dummies
978-0-7645-5193-2

Italian Phrases
For Dummies
978-0-7645-7203-6

Spanish For Dummies
978-0-7645-5194-9

Spanish For Dummies,
Audio Set
978-0-470-09585-0

Macintosh
Mac OS X Snow Leopard
For Dummies
978-0-470-43543-4

Math & Science
Algebra I For Dummies
978-0-7645-5325-7

Biology For Dummies
978-0-7645-5326-4

Calculus For Dummies
978-0-7645-2498-1

Chemistry For Dummies
978-0-7645-5430-8

Microsoft Office
Excel 2007 For Dummies
978-0-470-03737-9

Office 2007 All-in-One
Desk Reference
For Dummies
978-0-471-78279-7

Music
Guitar For Dummies,
2nd Edition
978-0-7645-9904-0

iPod & iTunes
For Dummies,
6th Edition
978-0-470-39062-7

Piano Exercises
For Dummies
978-0-470-38765-8

Parenting & Education
Parenting For Dummies,
2nd Edition
978-0-7645-5418-6

Type 1 Diabetes
For Dummies
978-0-470-17811-9

Pets
Cats For Dummies,
2nd Edition
978-0-7645-5275-5

Dog Training For Dummies,
2nd Edition
978-0-7645-8418-3

Puppies For Dummies,
2nd Edition
978-0-470-03717-1

Religion & Inspiration
The Bible For Dummies
978-0-7645-5296-0

Catholicism For Dummies
978-0-7645-5391-2

Women in the Bible
For Dummies
978-0-7645-8475-6

Self-Help & Relationship
Anger Management
For Dummies
978-0-470-03715-7

Overcoming Anxiety
For Dummies
978-0-7645-5447-6

Sports
Baseball For Dummies,
3rd Edition
978-0-7645-7537-2

Basketball For Dummies,
2nd Edition
978-0-7645-5248-9

Golf For Dummies,
3rd Edition
978-0-471-76871-5

Web Development
Web Design All-in-One
For Dummies
978-0-470-41796-6

Windows Vista
Windows Vista
For Dummies
978-0-471-75421-3